Transportation and Vehicular Engineering

Transportation and Vehicular Engineering

Edited by
Jeremiah Wallace

Larsen & Keller
www.larsen-keller.com

Transportation and Vehicular Engineering
Edited by Jeremiah Wallace
ISBN: 978-1-63549-278-1 (Hardback)

Larsen & Keller
Published by Larsen and Keller Education,
5 Penn Plaza,
19th Floor,
New York, NY 10001, USA

Cataloging-in-Publication Data

Transportation and vehicular engineering / edited by Jeremiah Wallace.
p. cm.
Includes bibliographical references and index.
ISBN 978-1-63549-278-1
1. Transportation engineering. 2. Transportation--Technological innovations. 3. Motor vehicles--Design and construction.
4. Automobiles--Design and construction. I. Wallace, Jeremiah.
TA1145 .T73 2017
629.04--dc23

Printed and bound in China.

For more information regarding Larsen and Keller Education and its products, please visit the publisher's website www.larsen-keller.com

Table of Contents

Preface

As a subfield of civil engineering, transportation engineering refers to the science and technology of the design, planning, operating and management facilities for any vehicle so as to provide a comfortable, safe, convenient, economic, efficient and environment friendly movement of goods and people. It also includes elements of urban planning. This book elucidates the concepts and innovative models around prospective developments with respect to transportation and vehicular engineering. The topics covered in this extensive book deal with the core subjects of this field. The various sub-fields along with technological progress that have future implications are also glanced at in it. Coherent flow of topics, student-friendly language and extensive use of examples make this textbook an invaluable source of knowledge.

Given below is the chapter wise description of the book:

Chapter 1- This chapter introduces the disciplines of transportation engineering, vehicle engineering and automotive engineering. These are the key branches that contribute in the manufacture of motor vehicles. The reader is provided with extensive information about each discipline and its various branches. The chapter provides an insightful focus, keeping in mind the complex subject matter.

Chapter 2- Transportation engineering is a vast discipline that incorporates functional design, planning, operations and the management of facilities of any mode of transportation. It is important as it ensures safety, efficiency, economy, environment compatibility etc. This chapter explores the categories of highway engineering and railway engineering that falls under transportation engineering and provides elaborate information and the discerning features of each branch.

Chapter 3- Transportation planning involves meticulous research on available alternates and the impact of the design, siting and policies regarding transportation. The chapter begins with a comprehensive study on the process of transportation planning. It then delves into aspects of transportation planning like streets, highway, bike lane, public transport and ramp meter. The key components of each aspect are discussed thoroughly in this chapter.

Chapter 4- Artificial intelligence, computer simulation and real-time information on traffic tendencies all provide smarter alternatives to the current scenario of traffic regulation and management. Intelligent transportation systems are a complex of intelligent systems that can process and implement efficient transportation and regulation of vehicular traffic. Transportation and vehicular engineering is best understood in confluence with the major topics listed in the following chapter.

Chapter 5- Transportation planning and engineering requires real-time movement of traffic and the use of transportation facilities for future road as well as urban transport and manufacture. It also prominently features in environmental conservation and urban pollution. The aspects elucidated in this chapter are of vital importance, and provide a better understanding of transportation and vehicular engineering.

Chapter 6- Traffic engineering is a branch of civil engineering; this subject focuses on the techniques used to achieve safe and efficient methods for road safety. This subject is closely related to transport engineering, pavement engineering, highway engineering and urban planning. This section helps the readers in developing an in-depth understanding of all the aspects of traffic engineering.

Indeed, my job was extremely crucial and challenging as I had to ensure that every chapter is informative and structured in a student-friendly manner. I am thankful for the support provided by my family and colleagues during the completion of this book.

Editor

1

Introduction to Transportation and Vehicle Engineering

This chapter introduces the disciplines of transportation engineering, vehicle engineering and automotive engineering. These are the key branches that contribute in the manufacture of motor vehicles. The reader is provided with extensive information about each discipline and its various branches. The chapter provides an insightful focus, keeping in mind the complex subject matter.

Transportation Engineering

Transportation engineering or transport engineering is the application of technology and scientific principles to the planning, functional design, operation and management of facilities for any mode of transportation in order to provide for the safe, efficient, rapid, comfortable, convenient, economical, and environmentally compatible movement of people and goods (transport). It is a sub-discipline of civil engineering and of industrial engineering. Transportation engineering is a major component of the civil engineering and mechanical engineering disciplines, according to specialisation of academic courses and main competences of the involved territory. The importance of transportation engineering within the civil and industrial engineering profession can be judged by the number of divisions in ASCE (American Society of Civil Engineers) that are directly related to transportation. There are six such divisions (Aerospace; Air Transportation; Highway; Pipeline; Waterway, Port, Coastal and Ocean; and Urban Transportation) representing one-third of the total 18 technical divisions within the ASCE (1987).

The planning aspects of transportation engineering relate to elements of urban planning, and involve technical forecasting decisions and political factors. Technical forecasting of passenger travel usually involves an urban transportation planning model, requiring the estimation of trip generation (how many trips for what purpose), trip distribution (destination choice, where is the traveler going), mode choice (what mode is being taken), and route assignment (which streets or routes are being used). More sophisticated forecasting can include other aspects of traveler decisions, including auto ownership, trip chaining (the decision to link individual trips together in a tour) and the choice of residential or business location (known as land use forecasting). Passenger trips are the focus of transportation engineering because they often represent the peak of demand on any transportation system.

A review of descriptions of the scope of various committees indicates that while facility planning and design continue to be the core of the transportation engineering field, such areas as operations planning, logistics, network analysis, financing, and policy analysis are also important to civil engineers, particularly to those working in highway and urban transportation. The National Council of Examiners for Engineering and Surveying (NCEES) list online the safety protocols, geometric design requirements, and signal timing.

Transportation engineering, as practiced by civil engineers, primarily involves planning, design, construction, maintenance, and operation of transportation facilities. The facilities support air, highway, railroad, pipeline, water, and even space transportation. The design aspects of transportation engineering include the sizing of transportation facilities (how many lanes or how much capacity the facility has), determining the materials and thickness used in pavement designing the geometry (vertical and horizontal alignment) of the roadway (or track).

Before any planning occurs the Engineer must take what is known as an inventory of the area or if it is appropriate, the previous system in place. This inventory or database must include information on (1)population, (2)land use, (3)economic activity, (4)transportation facilities and services, (5)travel patterns and volumes, (6)laws and ordinances, (7)regional financial resources, (8)community values and expectations. These inventories help the engineer create business models to complete accurate forecasts of the future conditions of the systemReview.

Operations and management involve traffic engineering, so that vehicles move smoothly on the road or track. Older techniques include signs, signals, markings, and tolling. Newer technologies involve intelligent transportation systems, including advanced traveler information systems (such as variable message signs), advanced traffic control systems (such as ramp meters), and vehicle infrastructure integration. Human factors are an aspect of transportation engineering, particularly concerning driver-vehicle interface and user interface of road signs, signals, and markings.

Highway Engineering

Engineers in this specialization:

- Handle the planning, design, construction, and operation of highways, roads, and other vehicular facilities as well as their related bicycle and pedestrian realms.
- Estimate the transportation needs of the public and then secure the funding for the project.
- Analyze locations of high traffic volumes and high collisions for safety and capacity.
- Use civil engineering principles to improve the transportation system.
- Utilizes the three design controls which are the drivers, the vehicles, and the roadways themselves.

Railroad Engineering

Railway engineers handle the design, construction, and operation of railroads and mass transit systems that use a fixed guideway (such as light rail or even monorails). Typical tasks would include determining horizontal and vertical alignment design, station location and design, and construction cost estimating. Railroad engineers can also move into the specialized field of train dispatching which focuses on train movement control.

Railway engineers also work to build a cleaner and safer transportation network by reinvesting and revitalizing the rail system to meet future demands. In the United States, railway engineers work with elected officials in Washington, D.C. on rail transportation issues to make sure that the rail system meets the country's transportation needs.

Port and Harbor Engineering

Port and harbor engineers handle the design, construction, and operation of ports, harbors, canals, and other maritime facilities.

Airport Engineering

Airport engineers design and construct airports. Airport engineers must account for the impacts and demands of aircraft in their design of airport facilities. These engineers must use the analysis of predominant wind direction to determine runway orientation, determine the size of runway border and safety areas, different wing tip to wing tip clearances for all gates and must designate the clear zones in the entire port.

Professional Societies

- Young Professionals in Transportation (YPT)
- Institute of Transportation Engineers (ITE)
- American Society for Highway Engineering (ASHE)
- American Society of Civil Engineers (ASCE)
- The American Association of State Highway and Transportation Officials (AASHTO)
- Eastern Asia Society for Transportation Studies (EASTS)

Vehicle Engineering

Vehicle engineering encompasses the fields of automotive engineering, aerospace engineering, rolling stock and marine engineering.

Automotive Engineering

Automotive engineering is the design, manufacture and operation of motorcycles, automobiles and trucks and their respective engineering subsystems. Automotive engineering is one of the most exciting professions you can choose. From the global concerns of sustainable mobility, and teaching cars to drive themselves, to working out how we'll get around on the surface of Mars, automotive engineering is all about the future. The work of an automobile down into three categories:

Design: Designing new products and improving existing ones

Research and Development: Finding solutions to engineering problems

Production: Planning and designing new production processes

Aerospace Engineering

Aerospace engineering is the branch of engineering that deals with the design, development, testing, and production of aircraft and related systems (aeronautical engineering) and of spacecraft, missiles, rocket-propulsion systems, and other equipment operating beyond the earth's atmosphere(astronautical engineering) Further concerned with the science of force and physics that are particular only to performance in Earth's atmosphere and the expanse of space.

Aeronautics

Definition

Aeronautics is the study of the science of flight. Aeronautics is the method of designing an airplane or other flying machine. There are four basic areas that aeronautical engineers must understand in order to be able to design planes. To design a plane, engineers must understand all of these elements.

Design Process

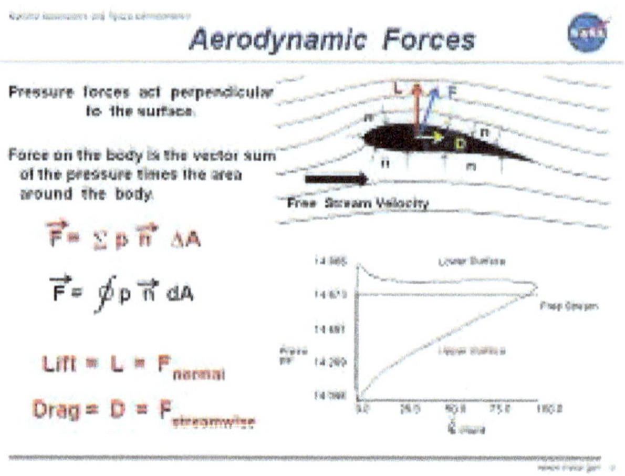

- Aerodynamics: is the study of how air flows around the airplane. By studying the way air flows around the plane the engineers can define the shape of the plane. The wings, the tail, and the main body or fuselage of the plane all affect the way the air will move around the plane.
- Propulsion: is the study of how to design an engine that will provide the thrust that is needed for a plane to take off and fly through the air. The engine provides the power for the airplane. The study of propulsion is what leads the engineers to determine the right kind of engine and the right amount of power that a plane will need
- Materials and Structures: is the study of what materials are to be used on the plane and in the engine and how those materials make the plane strong enough to fly effectively. The choice of materials that are used to make the fuselage wings, tail and engine will affect the strength and stability of the plane. Many airplane materials are now made out of composites, materials that are stronger than most metals and are lightweight.
- Stability and Control: is the study of how to control the speed, direction, altitude and other conditions that affect how a plane flies. The engineers design the controls that are needed in order to fly and instruments are provided for the pilot in the cockpit of the plane. The pilot uses these instruments to control the stability of the plane during flight.

Astronautics

Astronautics is the design and development of spacecraft with an emphasis on spacecraft systems, the design of ground control systems for spacecraft, and the design of orbital mechanics for spacecraft missions.

Rolling Stock

Rolling stock comprises all the vehicles that move on a railway. It usually includes both powered and unpowered vehicles, for example locomotives, railroad cars, coaches, and wagons.

Naval Architecture

Naval architecture also known as Naval engineering is an engineering discipline dealing with the design, construction, maintenance and operation of marine vessels and structures.

Automotive Engineering

Automotive engineering, along with aerospace engineering and marine engineering, is a branch of vehicle engineering, incorporating elements of mechanical, electrical, electronic, software and safety engineering as applied to the design, manufacture and operation of motorcycles, automobiles and trucks and their respective engineering subsystems.

Disciplines

Automotive Engineering

Some of the engineering attributes and disciplines that are of importance to the automotive engineer and many of the other aspects are included in it:

Safety engineering: Safety engineering is the assessment of various crash scenarios and their impact on the vehicle occupants. These are tested against very stringent governmental regulations. Some of these requirements include: seat belt and air bag functionality testing, front and side impact testing, and tests of rollover resistance. Assessments are done with various methods and tools, including Computer crash simulation (typically finite element analysis), crash test dummies, and partial system sled and full vehicle crashes.

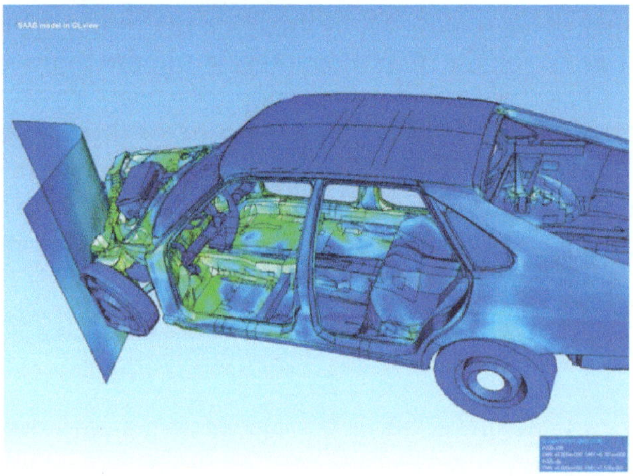

Visualization of how a car deforms in an asymmetrical crash using finite element analysis.

Fuel economy/emissions: Fuel economy is the measured fuel efficiency of the vehicle in miles per gallon or kilometers per litre. Emissions testing includes the measurement of vehicle emissions, including hydrocarbons, nitrogen oxides (NOx), carbon monoxide (CO), carbon dioxide (CO_2), and evaporative emissions.

Vehicle dynamics: Vehicle dynamics is the vehicle's response of the following attributes: ride, handling, steering, braking, comfort and traction. Design of the chassis systems of suspension, steering, braking, structure (frame), wheels and tires, and traction control are highly leveraged by the vehicle dynamics engineer to deliver the vehicle dynamics qualities desired.

NVH engineering (noise, vibration, and harshness): NVH is the customer's feedback (both tactile [felt] and audible [heard]) from the vehicle. While sound can be interpreted as a rattle, squeal, or hoot, a tactile response can be seat vibration, or a buzz in the steering wheel. This feedback is generated by components either rubbing, vibrating, or rotating. NVH response can be classified in various ways: powertrain NVH, road noise, wind noise, component noise, and squeak and rattle. Note, there are both good and bad NVH qualities. The NVH engineer works to either eliminate bad NVH, or change the "bad NVH" to good (i.e., exhaust tones).

Vehicle Electronics: Automotive electronics is an increasingly important aspect of automotive engineering. Modern vehicles employ dozens of electronic systems. These systems are responsible for operational controls such as the throttle, brake and steering controls; as well as many comfort and convenience systems such as the HVAC, infotainment, and lighting systems. It would not be possible for automobiles to meet modern safety and fuel economy requirements without electronic controls.

Performance: Performance is a measurable and testable value of a vehicles ability to perform in various conditions. Performance can be considered in a wide variety of tasks, but it's generally associated with how quickly a car can accelerate (e.g. standing start 1/4 mile elapsed time, 0–60 mph, etc.), its top speed, how short and quickly a car can come to a complete stop from a set speed (e.g. 70-0 mph), how much g-force a car can generate without losing grip, recorded lap times, cornering speed, brake fade, etc. Performance can also reflect the amount of control in inclement weather (snow, ice, rain).

Shift quality: Shift quality is the driver's perception of the vehicle to an automatic transmission shift event. This is influenced by the powertrain (engine, transmission), and the vehicle (driveline, suspension, engine and powertrain mounts, etc.) Shift feel is both a tactile (felt) and audible (heard) response of the vehicle. Shift quality is experienced as various events: Transmission shifts are felt as an upshift at acceleration (1–2), or a downshift maneuver in passing (4–2). Shift engagements of the vehicle are also evaluated, as in Park to Reverse, etc.

Durability / corrosion engineering: Durability and corrosion engineering is the evaluation testing of a vehicle for its useful life. Tests include mileage accumulation, severe driving conditions, and corrosive salt baths.

Package / ergonomics engineering: Package engineering is a discipline that designs/analyzes the occupant accommodations (seat roominess), ingress/egress to the vehicle, and the driver's field of vision (gauges and windows). The package engineer is also responsible for other areas of the vehicle like the engine compartment, and the component to component placement. Ergonomics is the discipline that assesses the occupant's access to the steering wheel, pedals, and other driver/passenger controls.

Climate control: Climate control is the customer's impression of the cabin environment and level of comfort related to the temperature and humidity. From the windshield defrosting, to the heating and cooling capacity, all vehicle seating positions are evaluated to a certain level of comfort.

Drivability: Drivability is the vehicle's response to general driving conditions. Cold starts and stalls, RPM dips, idle response, launch hesitations and stumbles, and performance levels.

Cost: The cost of a vehicle program is typically split into the effect on the variable cost of the vehicle, and the up-front tooling and fixed costs associated with developing the vehicle. There are also costs associated with warranty reductions, and marketing.

Program timing: To some extent programs are timed with respect to the market, and also to the production schedules of the assembly plants. Any new part in the design must support the development and manufacturing schedule of the model.

Assembly feasibility: It is easy to design a module that is hard to assemble, either resulting in damaged units, or poor tolerances. The skilled product development engineer works with the assembly/manufacturing engineers so that the resulting design is easy and cheap to make and assemble, as well as delivering appropriate functionality and appearance.

Quality management: Quality control is an important factor within the production process, as high quality is needed to meet customer requirements and to avoid expensive recall campaigns. The complexity of components involved in the production process requires a combination of different tools and techniques for quality control. Therefore, the International Automotive Task Force (IATF), a group of the world's leading manufacturers and trade organizations, developed the standard ISO/TS 16949. This standard defines the design, development, production, and when relevant, installation and service requirements. Furthermore, it combines the principles of ISO 9001 with aspects of various regional and national automotive standards such as AVSQ (Italy), EAQF (France), VDA6 (Germany) and QS-9000 (USA). In order to further minimize risks related to product failures and liability claims of automotive electric and electronic systems, the quality discipline functional safety according to ISO/IEC 17025 is applied.

Since the 1950s, the comprehensive business approach total quality management, TQM, helps to continuously improve the production process of automotive products and components. Some of the companies who have implemented TQM include Ford Motor Company, Motorola and Toyota Motor Company.

Job Functions

Development Engineer

A development engineer has the responsibility for coordinating delivery of the engineering attributes of a complete automobile (bus, car, truck, van, SUV, motorcycle etc.) as dictated by the automobile manufacturer, governmental regulations, and the customer who buys the product.

Much like the Systems Engineer, the development engineer is concerned with the interactions of all systems in the complete automobile. While there are multiple components and systems in an automobile that have to function as designed, they must also work in harmony with the complete automobile. As an example, the brake system's main function is to provide braking functionality to the automobile. Along with this, it must also provide an acceptable level of: pedal feel (spongy, stiff), brake system "noise" (squeal, shudder, etc.), and interaction with the ABS (anti-lock braking system)

Another aspect of the development engineer's job is a trade-off process required to deliver all of the automobile attributes at a certain acceptable level. An example of this is the trade-off between engine performance and fuel economy. While some customers are looking for maximum power from their engine, the automobile is still required to deliver an acceptable level of fuel economy. From the engine's perspective, these are opposing requirements. Engine performance is looking for maximum displacement (bigger, more power), while fuel economy is looking for a smaller displacement engine (ex: 1.4 L vs. 5.4 L). The engine size however, is not the only contributing factor to fuel economy and automobile performance.

Other attributes that involve trade-offs include: automobile weight, aerodynamic drag, transmission gearing, emission control devices, handling/roadholding, ride quality, and tires.

The development engineer is also responsible for organizing automobile level testing, validation, and certification. Components and systems are designed and tested individually by the Product Engineer. The final evaluation is to be conducted at the automobile level to evaluate system to system interactions. As an example, the audio system (radio) needs to be evaluated at the automobile level. Interaction with other electronic components can cause interference. Heat dissipation of the system and ergonomic placement of the controls need to be evaluated. Sound quality in all seating positions needs to be provided at acceptable levels.

Manufacturing Engineer

Manufacturing Engineers are responsible for ensuring proper production of the automotive components or complete vehicles. While the development engineers are responsible for the function of the vehicle, manufacturing engineers are responsible for the safe and effective production of the vehicle. This group of engineers consist of Process Engineers, Logistic Coordinators, Tooling Engineers, Robotics Engineers, and Assembly Planners.

In the automotive industry manufacturers are playing a larger role in the development stages of automotive components to ensure that the products are easy to manufacture. Design for Manufacturability in the automotive world is crucial to make certain whichever design is developed in the Research and Development Stage of automotive design. Once the design is established, the manufacturing engineers take over. They design the machinery and tooling necessary to build the automotive components or vehicle and establish the methods of how to mass-produce the product. It is the manufacturing engineers job to increase the efficiency of the automotive plant and to implement lean manufacturing techniques such as Six Sigma and Kaizen.

Other Automotive Engineering Roles

Other automotive engineers include those listed below:

- Aerodynamics engineers will often give guidance to the styling studio so that the shapes they design are aerodynamic, as well as attractive.
- Body engineers will also let the studio know if it is feasible to make the panels for their designs.
- Change control engineers make sure that all of the design and manufacturing changes that occur are organized, managed and implemented.
- Acoustics engineers are specific types of development engineers who do sound and aerodynamic testing to prevent loud cabin noises while the vehicle is on the road.

The Modern Automotive Product Engineering Process

Studies indicate that a substantial part of the modern vehicle's value comes from intelligent systems, and that these represent most of the current automotive innovation. To facilitate this, the modern automotive engineering process has to handle an increased use of mechatronics. Configuration and performance optimization, system integration, control, component, subsystem and system-level validation of the intelligent systems must become an intrinsic part of the standard vehicle engineering process, just as this is the case for the structural, vibro-acoustic and kinematic design. This requires a vehicle development process that is typically highly simulation-driven.

The V-approach

One way to effectively deal with the inherent multi-physics and the control systems development that is involved when including intelligent systems, is to adopt the V-Model approach to systems development, as has been widely used in the automotive industry for twenty years or more. In this V-approach, system-level requirements are propagated down the V via subsystems to component design, and the system performance is validated at increasing integration levels. Engineering of mechatronic systems requires the application of two interconnected "V-cycles": one focusing on the multi-physics system engineering (like the mechanical and electrical components of an electrically powered steering system, including sensors and actuators); and the other focuses on the controls engineering, the control logic, the software and realization of the control hardware and embedded software.

Predictive Engineering Analytics

An alternative approach is called predictive engineering analytics, and takes the V-approach to the next level. It lets design continue after product delivery. That is important for development of built-in predictive functionality and for creating vehicles that can be optimized while being in use, even based on real use data. This approach is based on the creation of a Digital Twin, a replica of the real product that remains in-sync. Manufacturers try to achieve this by implementing a set of development tactics and tools. Critical is a strong alignment of 1D systems simulation, 3D CAE an physical testing to reach more realism in the simulation process. This is combined with intelligent reporting and data analytics for better insight in the vehicle use. By supporting this with a strong data management structure that spans the entire product lifecycle, they bridge the gap between design, manufacturing and product use.

2

Various Aspects of Transportation Engineering

Transportation engineering is a vast discipline that incorporates functional design, planning, operations and the management of facilities of any mode of transportation. It is important as it ensures safety, efficiency, economy, environment compatibility etc. This chapter explores the categories of highway engineering and railway engineering that falls under transportation engineering and provides elaborate information and the discerning features of each branch.

Highway Engineering

Highway engineering is an engineering discipline branching from civil engineering that involves the planning, design, construction, operation, and maintenance of roads, bridges, and tunnels to ensure safe and effective transportation of people and goods. Highway engineering became prominent towards the latter half of the 20th Century after World War 2. Standards of highway engineering are continuously being improved. Highway engineers must take into account future traffic flows, design of highway intersections/interchanges, geometric alignment and design, highway pavement materials and design, structural design of pavement thickness, and pavement maintenance.

German Autobahn, 1936–1939

History

The beginning of road construction could be dated to the time of the Romans. With the advancement of technology from carriages pulled by two horses to vehicles with power equivalent to 100 horses, road development had to follow suit. The construction of modern highways did not begin until the late 19th to early 20th century.

The first research dedicated to highway engineering was initiated in the United Kingdom with the introduction of the Transport Research Laboratory (TRL), in 1930. In the USA, highway engineering became an important discipline with the passing of the Federal-Aid Highway Act of 1944, which aimed to connect 90% of cities with a population of 50,000 or more. With constant stress from vehicles which grew larger as time passed, improvements to pavement were needed. With technology out of date, in 1958 the construction of the first motorway in Great Britain (the Preston bypass) played a major role in the development of new pavement technology.

Design policies standards used in the United States are typically based on publications of the American Association of State Highway and Transportation Officials as well as research promulgated by the Transportation Research Board, the Institute of Transportation Engineers, the Federal Highway Administration, and the Department of Transportation.

Planning and Development

Highway planning involves the estimation of current and future traffic volumes on a road network. Highway engineers strive to predict and analyze all possible civil impacts of highway systems. Some considerations are the adverse effects on the environment, such as noise pollution, air pollution, water pollution, and other ecological impacts.

Financing

Developed countries are constantly faced with high maintenance cost of aging transportation highways. The growth of the motor vehicle industry and accompanying economic growth has generated a demand for safer, better performing, less congested highways. The growth of commerce, educational institutions, housing, and defense have largely drawn from government budgets in the past, making the financing of public highways a challenge.

The multipurpose characteristics of highways, economic environment, and the advances in highway pricing technology are constantly changing. Therefore, the approaches to highway financing, management, and maintenance are constantly changing as well.

Environmental Impact Assessment

The economic growth of a community is dependent upon highway development to enhance mobility. However, improperly planned, designed, constructed, and maintained highways can disrupt the social and economic characteristics of any size community. Common adverse impacts to highway development include damage of habitat and bio-diversity, creation of air and water pollution, noise/vibration generation, damage of natural landscape, and the destruction of a community's social and cultural structure. Highway infrastructure must be constructed and maintained to high qualities and standards.

There are three key steps for integrating environmental considerations into the planning, scheduling, construction, and maintenance of highways. This process is known as an Environmental Impact Assessment, or EIA, as it systematically deals with the following elements:

1. Identification of the full range of possible impacts on the natural and socio-economic environment

2. Evaluation and quantification of these impacts
3. Formulation of measures to avoid, mitigate, and compensate for the anticipated impacts.

Highway Safety

Highway systems generate the highest price in human injury and death, as nearly 50 million persons are injured in traffic accidents every year, not including the 1.2 million deaths. Road traffic injury is the single leading cause of unintentional death in the first five decades of human life.

Management of safety is a systematic process that strives to reduce the occurrence and severity of traffic accidents. The man/machine interaction with road traffic systems is unstable and poses a challenge to highway safety management. The key for increasing the safety of highway systems is to design, build, and maintain them to be far more tolerant of the average range of this man/machine interaction with highways. Technological advancements in highway engineering have improved the design, construction, and maintenance methods used over the years. These advancements have allowed for newer highway safety innovations.

By ensuring that all situations and opportunities are identified, considered, and implemented as appropriate, they can be evaluated in every phase of highway planning, design, construction, maintenance, and operation to increase the safety of our highway systems.

Design

The most appropriate location, alignment, and shape of a highway are selected during the design stage. Highway design involves the consideration of three major factors (human, vehicular, and roadway) and how these factors interact to provide a safe highway. Human factors include reaction time for braking and steering, visual acuity for traffic signs and signals, and car-following behaviour. Vehicle considerations include vehicle size and dynamics that are essential for determining lane width and maximum slopes, and for the selection of design vehicles. Highway engineers design road geometry to ensure stability of vehicles when negotiating curves and grades and to provide adequate sight distances for undertaking passing maneuvers along curves on two-lane, two-way roads.

Geometric Design

Highway and transportation engineers must meet many safety, service, and performance standards when designing highways for certain site topography. Highway geometric design primarily refers to the visible elements of the highways. Highway engineers who design the geometry of highways must also consider environmental and social effects of the design on the surrounding infrastructure.

There are certain considerations that must be properly addressed in the design process to successfully fit a highway to a site's topography and maintain its safety. Some of these design considerations include:

- Design speed
- Design traffic volume

- Number of lanes
- Level of Service (LOS)
- Sight Distance
- Alignment, super-elevation, and grades
- Cross section
- Lane width
- Horizontal and vertical clearance

The operational performance of a highway can be seen through drivers' reactions to the design considerations and their interaction.

Materials

The materials used for roadway construction have progressed with time, dating back to the early days of the Roman Empire. Advancements in methods with which these materials are characterized and applied to pavement structural design has accompanied this advancement in materials.

There are two major types of pavement surfaces - Portland cement concrete (PCC) and hot-mix asphalt (HMA). Underneath this wearing course are material layers that give structural support for the pavement system. These underlying surfaces may include either the aggregate base and subbase layers, or treated base and subbase layers, and additionally the underlying natural or treated subgrade. These treated layers may be cement-treated, asphalt-treated, or lime-treated for additional support.

Flexible Pavement Design

A flexible, or asphalt, or Tarmac pavement typically consists of three or four layers. For a four layer flexible pavement, there is a surface course, base course, and subbase course constructed over a compacted, natural soil subgrade. When building a three layer flexible pavement, the subbase layer is not used and the base course is placed directly on the natural subgrade.

A flexible pavement's surface layer is constructed of hot-mix asphalt (HMA).Unstabilized aggregates are typically used for the base course; however, the base course could also be stabilized with asphalt, Foamed Bitumen,<Roadstone Recycling> Portland cement, or another stabilizing agent. The subbase is generally constructed from local aggregate material, while the top of the subgrade is often stabilized with cement or lime.

With flexible pavement, the highest stress occurs at the surface and the stress decreases as the depth of the pavement increases. Therefore, the highest quality material needs to be used for the surface, while lower quality materials can be used as the depth of the pavement increases. The term "flexible" is used because of the asphalts ability to bend and deform slightly, then return to its original position as each traffic load is applied and removed. It is possible for these small deformations to become permanent, which can lead to rutting in the wheel path over an extended time.

The service life of a flexible pavement is typically designed in the range of 15 to 20 years. Required thicknesses of each layer of a flexible pavement vary widely depending on the materials used, magnitude, number of repetitions of traffic loads, environmental conditions, and the desired service life of the pavement. Factors such as these are taken into consideration during the design process so that the pavement will last for the designed life without excessive distresses.

Rigid Pavement Design

Rigid pavements are generally used in constructing airports and major highways, such as those in the interstate highway system. In addition, they commonly serve as heavy-duty industrial floor slabs, port and harbor yard pavements, and heavy-vehicle park or terminal pavements. Like flexible pavements, rigid highway pavements are designed as all-weather, long-lasting structures to serve modern day high-speed traffic. Offering high quality riding surfaces for safe vehicular travel, they function as structural layers to distribute vehicular wheel loads in such a manner that the induced stresses transmitted to the subgrade soil are of acceptable magnitudes.

Portland cement concrete (PCC) is the most common material used in the construction of rigid pavement slabs. The reason for its popularity is due to its availability and the economy. Rigid pavements must be designed to endure frequently repeated traffic loadings. The typical designed service life of a rigid pavement is between 30 and 40 years, lasting about twice as long as a flexible pavement.

One major design consideration of rigid pavements is reducing fatigue failure due to the repeated stresses of traffic. Fatigue failure is common among major roads because a typical highway will experience millions of wheel passes throughout its service life. In addition to design criteria such as traffic loadings, tensile stresses due to thermal energy must also be taken into consideration. As pavement design has progressed, many highway engineers have noted that thermally induced stresses in rigid pavements can be just as intense as those imposed by wheel loadings. Due to the relatively low tensile strength of concrete, thermal stresses are extremely important to the design considerations of rigid pavements.

Rigid pavements are generally constructed in three layers - a prepared subgrade, base or subbase, and a concrete slab. The concrete slab is constructed according to a designed choice of plan dimensions for the slab panels, directly influencing the intensity of thermal stresses occurring within the pavement. In addition to the slab panels, temperature reinforcements must be designed to control cracking behavior in the slab. Joint spacing is determined by the slab panel dimensions.

Three main types of concrete pavements commonly used are Jointed plain concrete pavement (JPCP), jointed reinforced concrete pavement (JRCP), and continuously reinforced concrete pavements (CRCP). JPCP's are constructed with contraction joints which direct the natural cracking of the pavement. These pavements do not use any reinforcing steel. JRCP's are constructed with both contraction joints and reinforcing steel to control the cracking of the pavement. High temperatures and moisture stresses within the pavement creates cracking, which the reinforcing steel holds tightly together. At transverse joints, dowel bars are typically placed to assist with transferring the load of the vehicle across the cracking. CRCP's solely rely on continuous reinforcing steel to hold the pavement's natural transverse cracks together. Prestressed concrete pavements have also been used in the construction of highways; however, they are not as common as the other three. Prestressed pavements allow for a thinner slab thickness by partly or wholly neutralizing thermally induced stresses or loadings.

Flexible Pavement Overlay Design

Over the service life of a flexible pavement, accumulated traffic loads may cause excessive rutting or cracking, inadequate ride quality, or an inadequate skid resistance. These problems can be avoided by adequately maintaining the pavement, but the solution usually has excessive maintenance costs, or the pavement may have an inadequate structural capacity for the projected traffic loads.

Throughout a highway's life, its level of serviceability is closely monitored and maintained. One common method used to maintain a highway's level of serviceability is to place an overlay on the pavement's surface.

There are three general types of overlay used on flexible pavements: asphalt-concrete overlay, Portland cement concrete overlay, and ultra-thin Portland cement concrete overlay. The concrete layer in a conventional PCC overlay is placed unbonded on top of the flexible surface. The typical thickness of an ultra-thin PCC overlay is 4 inches (10 cm) or less.

There are three main categories of flexible pavement overlay design procedures:

1. Component Analysis Design
2. Deflection-Based Design

Rigid Pavement Overlay Design

Near the end of a rigid pavement's service life, a decision must be made to either fully reconstruct the worn pavement, or construct an overlay layer. Considering an overlay can be constructed on a rigid pavement that has not reached the end of its service life, it is often more economically attractive to apply overlay layers more frequently. The required overlay thickness for a structurally sound rigid pavement is much smaller than for one that has reached the end of its service life. Rigid and flexible overlays are both used for rehabilitation of rigid pavements such as JPCP, JRCP, and CRCP.

There are three subcategories of rigid pavement overlays that are organized depending on the bonding condition at the pavement overlay and existing slab interface.

1. Bonded overlays
2. Unbonded overlays
3. Partially bonded overlays

Drainage System Design

Designing for proper drainage of highway systems is crucial to their success. Regardless of how well other aspects of a road are designed and constructed, adequate drainage is mandatory for a road to survive its entire service life. Excess water in the highway structure can inevitably lead to premature failure, even if the failure is not catastrophic.

Each highway drainage system is site-specific and can be very complex. Depending on the geography of the region, many methods for proper drainage may not be applicable. The highway engineer must determine which situations a particular design process should be applied, usually a combination of several appropriate methods and materials to direct water away from the structure.

Erosion control is a crucial component in the design of highway drainage systems. Surface drainage must be allowed for precipitation to drain away from the structure. Highways must be designed with a slope or crown so that runoff water will be directed to the shoulder of the road, into a ditch, and away from the site. Designing a drainage system requires the prediction of runoff and infiltration, open channel analysis, and culvert design for directing surface water to an appropriate location.

Construction, Maintenance, and Management

Highway Construction

Highway construction is generally preceded by detailed surveys and subgrade preparation. The methods and technology for constructing highways has evolved over time and become increasingly sophisticated. This advancement in technology has raised the level of skill sets required to manage highway construction projects. This skill varies from project to project, depending on factors such as the project's complexity and nature, the contrasts between new construction and reconstruction, and differences between urban region and rural region projects.

There are a number of elements of highway construction which can be broken up into technical and commercial elements of the system. Some examples of each are listed below:

- **Technical Elements**
  - Materials
  - Material quality
  - Installation techniques
  - Traffic
- **Commercial Elements**
  - Contract understanding
  - Environmental aspects
  - Political aspects
  - Legal aspects
  - Public concerns

Typically, construction begins at the lowest elevation of the site, regardless of the project type, and moves upward. By reviewing the geotechnical specifications of the project, information is given about:

- Existing ground conditions
- Required equipment for excavation, grading, and material transportation to and from the site
- Properties of materials to be excavated
- Dewatering requirements necessary for below-grade work
- Shoring requirements for excavation protection
- Water quantities for compaction and dust control

Subbase Course Construction

A subbase course is a layer designed of carefully selected materials that is located between the subgrade and base course of the pavement. The subbase thickness is generally in the range of 4 to 16 inches, and it is designed to withstand the required structural capacity of the pavement section.

Common materials used for a highway subbase include gravel, crushed stone, or subgrade soil that is stabilized with cement, fly ash, or lime. Permeable subbase courses are becoming more prevalent because of their ability to drain infiltrating water from the surface. They also prevent subsurface water from reaching the pavement surface.

When local material costs are excessively expensively or the material requirements to increase the structural bearing of the sub-base are not readily available, highway engineers can increase the bearing capacity of the underlying soil by mixing in Portland cement, foamed asphalt, or with emerging technologies such as the cross-linking styrene acrylic polymer that increases the California Bearing Ratio of in-situ materials by a factor 4 – 6.

Base Course Construction

The base course is the region of the pavement section that is located directly under the surface course. If there is a subbase course, the base course is constructed directly about this layer. Otherwise, it is built directly on top of the subgrade. Typical base course thickness ranges from 4 to 6 inches and is governed by underlying layer properties.

Heavy loads are continuously applied to pavement surfaces, and the base layer absorbs the majority of these stresses. Generally, the base course is constructed with an untreated crushed aggregate such as crushed stone, slag, or gravel. The base course material will have stability under the construction traffic and good drainage characteristics.

The base course materials are often treated with cement, bitumen, calcium chloride, sodium chloride, fly ash, or lime. These treatments provide improved support for heavy loads, frost susceptibility, and serves as a moisture barrier between the base and surface layers.

Surface Course Construction

There are two most commonly used types of pavement surfaces used in highway construction: hot-mix asphalt and Portland cement concrete. These pavement surface courses provide a smooth and safe riding surface, while simultaneously transferring the heavy traffic loads through the various base courses and into the underlying subgrade soils.

Hot-mix Asphalt (HMA) Layers

Hot-mix asphalt surface courses are referred to as flexible pavements. The Superpave System was developed in the late 1980s and has offered changes to the design approach, mix design, specifications, and quality testing of materials.

The construction of an effective, long-lasting asphalt pavement requires an experienced construction crew, committed to their work quality and equipment control.

Construction Issues:

- Asphalt mix segregation
- Laydown
- Compaction
- Joints

A prime coat is a low viscosity asphalt that is applied to the base course prior to laying the HMA surface course. This coat bonds loose material, creating a cohesive layer between the base course and asphalt surface.

A tack coat is a low viscosity asphalt emulsion that is used to create a bond between an existing pavement surface and new asphalt overlay. Tack coats are typically applied on adjacent pavements (curbs) to assist the bonding of the HMA and concrete.

Portland Cement Concrete (PCC)

Portland cement concrete surface courses are referred to as rigid pavements, or concrete pavements. There are three general classifications of concrete pavements - jointed plain, jointed reinforced, and continuously reinforced.

Traffic loadings are transferred between sections when larger aggregates in the PCC mix inter-lock together, or through load transfer devices in the transverse joints of the surface. Dowel bars are used as load-transferring devices to efficiently transfer loads across transverse joints while maintaining the joint's horizontal and vertical alignment. Tie-bars are deformed steel bars that are placed along longitudinal joints to hold adjacent pavement sections in place.

Highway Maintenance

The overall purpose of highway maintenance is to fix defects and preserve the pavement's structure and serviceability. Defects must be defined, understood, and recorded in order to select an appropriate maintenance plan. Defects differ between flexible and rigid pavements.

There are four main objectives of highway maintenance:

- repair of functional pavement defects
- extend the functional and structural service life of the pavement
- maintain road safety and signage
- keep road reserve in acceptable condition

Through routine maintenance practices, highway systems and all of their components can be maintained to their original, as-built condition.

Project Management

Project management involves the organization and structuring of project activities from inception to completion. Activities could be the construction of infrastructure such as highways and bridges or major and minor maintenance activities related to constructing such infrastructure. The entire project and involved activities must be handled in a professional manner and completed within deadlines and budget. In addition, minimizing social and environmental impacts is essential to successful project management.

Railway Engineering

Railway engineering is a multi-faceted engineering discipline dealing with the design, construction and operation of all types of railway systems. It encompasses a wide range of engineering disciplines, including civil engineering, computer engineering, electrical engineering, mechanical engineering, industrial engineering and production engineering. A great many other engineering sub-disciplines are also called upon.

History

With the advent of the railways in the early nineteenth century, a need arose for a specialized group of engineers capable of dealing with the unique problems associated with railway engineering. As the railways expanded and became a major economic force, a great many engineers became involved in the field, probably the most notable in Britain being Richard Trevithick, George Stephenson and Isambard Kingdom Brunel. Today, railway systems engineering continues to be a vibrant field of engineering.

Subfields

- Mechanical engineering
- Command, control & railway signalling
    - Office systems design
    - Data center design
    - SCADA

    - Network design
- Electrical engineering
    - Energy electrification
        - Third rail
        - Overhead contact system
- Civil engineering
    - Permanent way engineering
- Light rail systems
- On-track plant
- Rail systems integration
- Train control systems
    - Cab signalling
- Railway vehicle engineering
- Rolling resistance
    - Curve resistance
- Wheel–rail interface
    - Hunting oscillation
- Railway systems engineering
    - Railway signalling
    - Fare collection
    - CCTV
    - Public address
    - Intrusion detection
    - Access control
- Systems integration

Professional Organizations

In the UK: The Railway Division of the Institution of Mechanical Engineers (IMechE).

In the US The American Railway Engineering and Maintenance-of-Way Association (AREMA)

Worldwide The Institute of Railway Signal Engineers (IRSE)

References

- Rogers, Martin (2002). Highway engineering. Oxford, UK: Blackwell Science. ISBN 978-0-632-05993-5.
- O'Flaherty, edited by C.A. (2002). Highways the location, design, construction and maintenance of road pavements (4th ed.). Oxford: Butterworth-Heinemann. ISBN 978-0-7506-5090-8.
- Highway engineering." McGraw-Hill Concise Encyclopedia of Science and Technology. New York: McGraw-Hill, 2006. Credo Reference. Web. 13 February 2013.

3

Understanding Transportation Planning

Transportation planning involves meticulous research on available alternates and the impact of the design, siting and policies regarding transportation. The chapter begins with a comprehensive study on the process of transportation planning. It then delves into aspects of transportation planning like streets, highway, bike lane, public transport and ramp meter. The key components of each aspect are discussed thoroughly in this chapter.

Transportation Planning

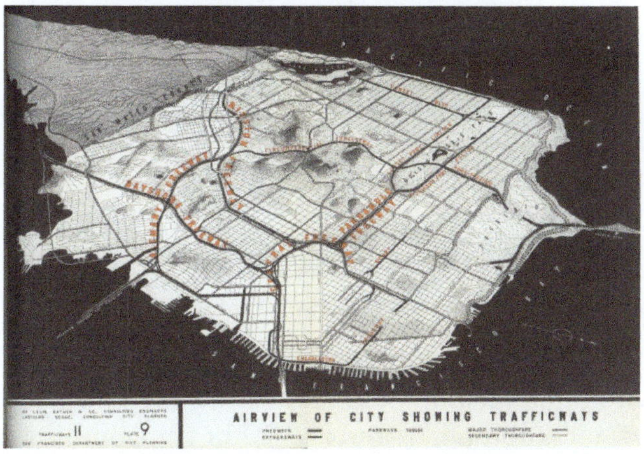

1948 San Francisco roadway plan

Transportation planning is the process of defining future policies, goals, investments and designs to prepare for future needs to move people and goods to destinations. As practiced today, it is a collaborative process that incorporates the input of many stakeholders including various government agencies, the public and private businesses. Transportation planners apply a multi-modal and/or comprehensive approach to analyzing the wide range of alternatives and impacts on the transportation system to influence beneficial outcomes.

Transportation planning is also commonly referred to as transport planning internationally, and is involved with the evaluation, assessment, design and siting of transport facilities (generally streets, highways, bike lanes and public transport lines).

Models and Sustainability

Transportation planning, or transport planning, has historically followed the rational planning model of defining goals and objectives, identifying problems, generating alternatives, evaluating alternatives, and developing plans. Other models for planning include rational actor, transit oriented development, satisficing, incremental planning, organizational process, and political bargaining.

Chicago Transit Authority Chicago 'L' trains use elevated tracks for a portion of the system known as the Loop, which is located in the Chicago Loop community area. This is an example of the siting of transportation facilities resulting from transportation planning.

A bypass the Old Town in Szczecin, Poland

Planners are increasingly expected to adopt a multi-disciplinary approach, especially due to the rising importance of environmentalism. For example, the use of behavioural psychology to persuade drivers to abandon their automobiles and use public transport instead. The role of the transport planner is shifting from technical analysis to promoting sustainability through integrated transport policies. For example, in Hanoi the increasing number of motorcycles is responsible not only for environmental damage but also for slowing down economic growth. In the long run the plan is to reduce traffic through a change in urban planning. Through economic incentives and attractive alternatives experts hope to lighten traffic in the short run.

United Kingdom

In the United Kingdom transport planning has traditionally been a branch of civil engineering. In the 1950s and 1960s it was generally believed that the motor car was an important element in the future of transport as economic growth spurred on car ownership figures. The role of the transport planner was to match motorway and rural road capacity against the demands of economic growth. Urban areas would need to be redesigned for the motor vehicle or else impose traffic containment and demand management to mitigate congestion and environmental impacts. These policies were popularised in a 1963 government publication, *Traffic in Towns*. The contemporary Smeed Report on congestion pricing was initially promoted to manage demand but was deemed politically unacceptable. In more recent times this approach has been caricatured as "predict and provide" – to predict future transport demand and provide the network for it, usually by building more roads.

The publication of Planning Policy Guidance 13 in 1994 (revised in 2001), followed by A New Deal for Transport in 1998 and the white paper Transport Ten Year Plan 2000 again indicated an acceptance that unrestrained growth in road traffic was neither desirable nor feasible. The worries were threefold: concerns about congestion, concerns about the effect of road traffic on the environment (both natural and built) and concerns that an emphasis on road transport discriminates against vulnerable groups in society such as the poor, the elderly and the disabled.

These documents reiterated the emphasis on integration:

- integration within and between different modes of transport
- integration with the environment
- integration with land use planning
- integration with policies for education, health and wealth creation.

This attempt to reverse decades of underinvestment in the transport system has resulted in a severe shortage of transport planners. It was estimated in 2003 that 2,000 new planners would be required by 2010 to avoid jeopardising the success of the Transport Ten Year Plan .

During 2006 the Transport Planning Society defined the key purpose of transport planning as:

> *to plan, design, deliver, manage and review transport, balancing the needs of society, the economy and the environment.*

The following key roles must be performed by transport planners:

- take account of the social, economic and environmental context of their work
- understand the legal, regulatory policy and resource framework within which they work
- understand and create transport policies, strategies and plans that contribute to meeting social, economic and environmental needs
- design the necessary transport projects, systems and services
- understand the commercial aspects of operating transport systems and services
- know about and apply the relevant tools and techniques
- must be competent in all aspects of management, in particular communications, personal skills and project management.

The UK Treasury recognises and has published guidance on the systematic tendency for project appraisers to be overly optimistic in their initial estimates.

United States

Transportation planning in the United States is in the midst of a shift similar to that taking place in the United Kingdom, away from the singular goal of moving vehicular traffic and towards an approach that takes into consideration the communities and lands through which streets, roads, and highways pass ("the context"). More so, it places a greater emphasis on passenger rail networks, which had been neglected until recently. This new approach, known as Context Sensitive Solutions (CSS), seeks to balance the need to move people efficiently and safely with other desirable outcomes, including historic preservation, environmental sustainability, and the creation of vital public spaces.

A Civilian Conservation Corps (CCC) map of the planned route of a parkway. During the 1930s, the CCC was actively involved in creating and improving roads throughout rural areas and parks

The initial guiding principles of CSS came out of the 1998 "Thinking Beyond the Pavement" conference as a means to describe and foster transportation projects that preserve and enhance the natural and built environments, as well as the economic and social assets of the neighborhoods they pass through. CSS principles have since been adopted as guidelines for highway design in federal legislation. Also, in 2003, the Federal Highway Administration announced that under one of its three Vital Few Objectives (Environmental Stewardship and Streamlining) they set the target of achieving CSS integration within all state Departments of Transportation by September 2007. The recent pushes for advancing transportation planning has led to the development of a professional certification program by the Institute of Transportation Engineers, the Professional Transportation Planner, to be launched in 2007. An advanced form of certification - the Advanced Specialty Certification in Transportation Planning was developed by the American Planning Association thereafter in 2011. The Certified Transportation Planner credential is only available for those professional planners (AICP members) whom have at a minimum of eight years of transportation planning experience.

In recent years, there has been a movement to provide "complete" transportation corridors under the "complete streets" movement. In response to auto-centric design of transportation networks, complete streets encompass all users and modes of transportation in a more equitable manner. The complete streets movement entails many of the CSS principles as well as pedestrian, bicycle and older adult movements to improve transportation in the United States.

The Technical Process

Most regional transport planners employ what is called the rational model of planning. The model views planning as a logical and technical process that uses the analysis of quantitative data to decide how to best invest resources in new and existing transport infrastructure.

Since World War II, this attitude in planning has resulted in the widespread use of travel modelling as a key component of regional transport planning. The models' rise in popularity can also be attributed to a rapid increase in the number of automobiles on the road, widespread suburbanization and a large increase in federal or national government spending upon transport in urban areas. All of these phenomena dominated the planning culture in the late 1940s, 1950s and 1960s. Regional transport planning was needed because increasingly cities weren't just cities anymore, but parts of a complex regional system.

The US process, according to Johnston (2004) and the FHWA and Federal Transit Administration (FTA) (2007), generally follows a pattern which can be divided into three different stages. Over the course of each of three phases, the metropolitan planning organization (MPO) is also supposed to consider air quality and environmental issues, look at planning questions in fiscally constrained way and involve the public. In the first stage, called preanalysis, the MPO considers what problems and issues the region faces and what goals and objectives it can set to help address those issues. During this phase the MPO also collects data on wide variety of regional characteristics, develops a set of different alternatives that will be explored as part of the planning process and creates a list of measurable outcomes that will be used to see whether goals and objectives have been achieved. Johnston notes that many MPOs perform weakly in this area, and though many of these activities seem like the "soft" aspects of planning that aren't really necessary, they're absolutely essential to ensuring that the models used in second phase are accurate and complete .

The second phase is technical analysis. The process involves much technical maneuvering, but basically the development of the models can be broken down as follows. Before beginning, the MPO collects enormous amounts of data. This data can be thought of as falling into two categories: data about the transport system and data about adjacent land use. The best MPOs are constantly collecting this data.

The actual analysis tool used in the US is called the Urban Transportation Modeling System (UTMS), though it is often referred to as the four-step process. As its nickname suggestions, UTMS has four steps: trip generation, trip distribution, mode choice and trip/route assignment. In trip generation, the region is subdivided into a large number of smaller units of analysis called traffic analysis zones (TAZs). Based on the number and characteristics of the households in each zone, a certain number of trips is generated. In the second step, trip distribution, trips are separated out into categories based on their origin and purpose: generally, these categories are home-based work, home-based other and non-home based. In each of three categories, trips are matched to origin and destination zones using the data that has been collected.

In mode choice, trips are assigned to a mode (usually auto or transit) based on what's available in a particular zone, the characteristics of the household within that zone and the cost of the mode for each mode in terms of money and time. Since most trips by bicycle or walking are generally shorter, they are assumed to have stayed within one zone and are not included in the analysis. Finally, in route assignment, trips are assigned to the network. As particular parts of the network are assigned trips, the vehicle speed slows down, so some trips are assigned to alternate routes in such a way that all trip times are equal. This is important because the ultimate goal is system-wide optimization, not optimization for any one individual. The finished product is traffic flows and speeds for each link in the network.

Ideally, these models would include all the different behaviours that are associated with transport, including complex policy questions which are more qualitative in nature. Because of the complexity of transport issues, this is often not possible in practice. This results in models which may estimate future traffic conditions well, but are ultimately based on assumptions made on the part of the planner. Some planners carry out additional sub-system modelling on things like automobile ownership, time of travel, location of land development, location and firms and location of households to help to fill these knowledge gaps, but what are created are nevertheless models, and models always include some level of uncertainty.

The post-analysis phase involves plan evaluation, programme implementation and monitoring of the results. Johnston notes that for evaluation to be meaningful it should be as comprehensive as possible. For example, rather than just looking at decreases in congestion, MPOs should consider economic, equity and environmental issues.

Politics and Transportation Planning

Although a transportation planning process may appear to be a rational process based on standard and objective methodologies, it is often influenced by political processes. Transportation planning is closely interrelated to the public nature of government works projects. As a result, transportation planners play both a technical and a coordinating role. Politicians often have vastly differing perspectives, goals and policy desires. Transportation planners help by providing information to decision makers, such as politicians, in a manner that produces beneficial outcomes. This role is similar to transportation engineers, who are often equally influenced by politics in the technical process of transportation engineering design.

Integration with Urban Planning

Transport isochrone maps can be used in connection with urban planning to evaluate sites.

Street

Typical service street ("mews") in the Royal Borough of Kensington and Chelsea in London. Mews are typically found at the back of older housing terraces (rows of townhouses) in the UK, with a more elegant street in front of the terrace.

A street is a public thoroughfare (usually paved) in a built environment. It is a public parcel of land adjoining buildings in an urban context, on which people may freely assemble, interact, and move about. A street can be as simple as a level patch of dirt, but is more often paved with a hard, durable surface such as concrete, cobblestone or brick. Portions may also be smoothed with asphalt, embedded with rails, or otherwise prepared to accommodate non-pedestrian traffic.

Abbey Road, London

Originally the word "street" simply meant a paved road (Latin: "via strata"). The word "street" is still sometimes used colloquially as a synonym for "road", for example in connection with the ancient Watling Street, but city residents and urban planners draw a crucial modern distinction: a road's main function is transportation, while streets facilitate public interaction. Examples of streets include pedestrian streets, alleys, and city-centre streets too crowded for road vehicles to pass. Conversely, highways and motorways are types of roads, but few would refer to them as streets.

Etymology

A city-centre street in Frankfurt, Germany.

The word *street* has its origins in the Latin *strata* (meaning "paved road" - abbreviation from *via strata*); it is thus related to *stratum* and *stratification*. Ancient Greek *stratos* means army; Greeks originally built roads to move their armies. Old English applied the word to Roman roads in Britain such as Ermine Street, Watling Street, etc. Later it acquired a dialectical meaning of "straggling village", which were often laid out on the verges of Roman roads and these settlements often became named Stretton. In the Middle Ages, a *road* was a way people travelled, with *street* applied specifically to paved ways.

The Porta Rosa was the main street of Elea, connecting the northern quarter with the southern quarter. The street is 5 meters wide and has an incline of 18 % in the steepest part. It is paved with limestone blocks, griders cut in square blocks, and on one side a small gutter for the drainage of rain water. The building is dated during the time of the reorganization of the city during the Hellenistic age (4th–3rd centuries BC).

Role in the Built Environment

Rue Saint-Jacques, a street in Montreal, 1910

The street is a public easement, one of the few shared between all sorts of people. As a component of the built environment as ancient as human habitation, the street sustains a range of activities vital to civilization. Its roles are as numerous and diverse as its ever-changing cast of characters.

Streets can be loosely categorized as main streets and side streets. Main streets are usually broad with a relatively high level of activity. Commerce and public interaction are more visible on main streets, and vehicles may use them for longer-distance travel. Side streets are quieter, often residential in use and character, and may be used for vehicular parking.

Circulation

Circulation, or less broadly, transportation, is perhaps a street's most visible use, and certainly among the most important. The unrestricted movement of people and goods within a city is essential to its commerce and vitality, and streets provide the physical space for this activity.

In the interest of order and efficiency, an effort may be made to segregate different types of traffic. This is usually done by carving a road through the middle for motorists, reserving pavements on either side for pedestrians; other arrangements allow for streetcars, trolleys, and even wastewater and rainfall runoff ditches (common in Japan and India). In the mid-20th century, as the automobile threatened to overwhelm city streets with pollution and ghastly accidents, many urban theorists came to see this segregation as not only helpful but necessary in order to maintain mobility.

Le Corbusier, for one, perceived an ever-stricter segregation of traffic as an essential affirmation of social order—a desirable, and ultimately inevitable, expression of modernity. To this end, proposals were advanced to build "vertical streets" where road vehicles, pedestrians, and trains would each occupy their own levels. Such an arrangement, it was said, would allow for even denser development in the future.

These plans were never implemented comprehensively, a fact which today's urban theorists regard as fortunate for vitality and diversity. Rather, vertical segregation is applied on a piecemeal basis, as in sewers, utility poles, depressed highways, elevated railways, common utility ducts, the extensive complex of underground malls surrounding Tokyo Station and the Ōtemachi subway station, the elevated pedestrian skyway networks of Minneapolis and Calgary, the underground cities of Atlanta and Montreal, and the multilevel streets in Chicago.

Transportation is often misunderstood to be the defining characteristic, or even the sole purpose, of a street. This has not been the case since the word "street" came to be limited to urban situations, and even in the automobile age, is still demonstrably false. A street may be temporarily blocked to all through traffic in order to secure the space for other uses, such as a street fair, a flea market, children at play, filming a movie, or construction work. Many streets are bracketed by bollards or Jersey barriers so as to keep out vehicles. These measures are often taken in a city's busiest areas, the "destination" districts, when the volume of activity outgrows the capacity of private passenger vehicles to support it. A feature universal to all streets is a human-scale design that gives its users the space and security to feel engaged in their surroundings, whatever through traffic may pass.

Vehicular Traffic

Despite this, the operator of a motor vehicle may (incompletely) regard a street as merely a thoroughfare for vehicular travel or parking. As far as concerns the driver, a street can be one-way or two-way: vehicles on one-way streets may travel in only one direction, while those on two-way streets may travel both ways. One way streets typically have signs reading "ONE WAY" and an arrow showing the direction of allowed travel. Most two-way streets are wide enough for at least two lanes of traffic.

A street full of vehicles in Shanghai

Kitano Street in Kobe, Hyogo, Japan.

An empty street in Misasa, Tottori, Japan.

Which lane is for which direction of traffic depends on what country the street is located in. On broader two-way streets, there is often a *centre line* marked down the middle of the street separating those lanes on which vehicular traffic goes in one direction from other lanes in which traffic goes in the opposite direction. Occasionally, there may be a median strip separating lanes of opposing traffic. If there is more than one lane going in one direction on a main street, these lanes may be separated by intermittent *lane lines* marked on the street pavement. Side streets often do not have centre lines or lane lines.

Parking for Vehicles

Many streets, especially side streets in residential areas, have an extra lane's width on one or both sides for parallel parking. Most minor side streets allowing free parallel parking do not have pavement markings designating the parking lane. Main streets more often have parking lanes marked. Some streets are too busy or narrow for parking on the side. Sometimes parking on the sides of streets is allowed only at certain times. Curbside signs often state regulations about parking. Some streets, particularly in business areas, may have parking meters into which coins must be paid to allow parking in the adjacent space for a limited time. Other parking meters work on a credit card and ticket basis or pay and display. Parking lane markings on the pavement may designate the meter corresponding to a parking space. Some wide streets with light traffic allow *angle parking*.

Sidewalk and Bicycle Traffic

Sidewalks (US usage) or pavements (UK usage) are often located alongside on one or usually both sides of the street within the public land strips beyond the curbs. Sidewalks serve a traffic purpose, by making walking easier and more attractive, but they also serve a social function, allowing neighbors to meet and interact on their walks. They also can foster economic activity, such as window shopping and sidewalk cafes. Some studies have found that shops on streets with sidewalks get more customers than similar shops without sidewalks.

An important element of sidewalk design is accessibility for persons with disabilities. Features that make sidewalks more accessible include curb ramps, tactile paving and accessible traffic signals. The Americans with Disabilities Act requires accessibility improvement on new and reconstructed streets within the US.

In most jurisdictions, bicycles are legally allowed to use streets, and required to follow the same traffic laws as motor vehicle traffic. Where the volume of bicycle traffic warrants and available right-of-way allows, provisions may be made to separate bicyclists from motor vehicle traffic. Wider lanes may be provided next to the curb, or shoulders may be provided. Bicycle lanes may be used on busy streets to provide some separation between bicycle traffic and motor vehicle traffic. The bicycle lane may be placed between the travel lanes and the parking lanes, between the parking lanes and the curb, or for increased safety for cyclists, between curb and sidewalk.

Vehicular Amenities and Roadside Hardware

Often, a curb is used to separate the vehicle traffic lanes from the adjacent sidewalk area. Street signs, parking meters, bicycle stands, benches, traffic signals, and street lights are often found adjacent to streets. They may be behind the sidewalk, or between the sidewalk and the curb.

Landscaping

There may be a road verge (a strip of grass or other vegetation) between the carriageway (North American English: Roadway) and the pavement on either side of the street on which Grass or trees are often grown there for landscaping. These are often placed for beautification, but are increasingly being used to control stormwater.

Utilities

Although primarily used for traffic, streets are important corridors for utilities such as electric power; communications such as telephone, cable television and fiber optic lines; storm and sanitary sewers; and natural gas lines.

Street Numbering

Practically all public streets in Western countries and the majority elsewhere are given a street or road name, or at least a number, to identify them and any addresses located along the streets. Alleys, in some places, do not have names. The length of a lot of land along a street is referred to as the *frontage* of the lot.

Interaction

Pedestrians walking along Elfreth's Alley, Philadelphia

A street may assume the role of a town square for its regulars. Jane Jacobs, an economist and prominent urbanist, wrote extensively on the ways that interaction among the people who live and work on a particular street—"eyes on the street"—can reduce crime, encourage the exchange of ideas, and generally make the world a better place.

Identity

A street can often serve as the catalyst for the neighborhood's prosperity, culture and solidarity. New Orleans' Bourbon Street is famous not only for its active nightlife but also for its role as the center of the city's French Quarter. Similarly, the Bowery has at various times been New York City's theater district, red-light district, skid row, restaurant supply district, and the center of the nation's underground punk scene. Madison Avenue and Fleet Street are so strongly identified with their respective most famous types of commerce, that their names are sometimes applied to firms located elsewhere. Other streets mark divisions between neighborhoods of a city. For example, Yonge Street divides Toronto into east and west sides, and East Capitol Street divides Washington, D.C. into north and south.

Some streets are associated with the beautification of a town or city. Greenwood, Mississippi's Grand Boulevard was once named one of America's ten most beautiful streets by the U.S. Chambers of Commerce and the Garden Clubs of America. The 1,000 oak trees lining Grand Boulevard were planted in 1916 by Sally Humphreys Gwin, a charter member of the Greenwood Garden Club. In 1950, Gwin received a citation from the National Congress of the Daughters of the American Revolution in recognition of her work in the conservation of trees.

Streets also tend to aggregate establishments of similar nature and character. East 9th Street in Manhattan, for example, offers a cluster of Japanese restaurants, clothing stores, and cultural venues. In Washington, D.C., 17th Street and P Street are well known as epicenters of the city's (relatively small) gay culture. Many cities have a Radio Row or Restaurant Row. Like in Philadelphia there is a small street called Jewelers' row giving the identity of a "Diamond district". This phenomenon is the subject of urban location theory in economics. In Cleveland, Ohio, East 4th Street has become restaurant row for Cleveland. On East 4th is Michael Symon's Lola Bistro and other restaurants.

As Distinct from Other Spaces

Centre Ville, Beirut, Lebanon

A road, like a street, is often paved and used for travel. However, a street is characterized by the degree and quality of street life it facilitates, whereas a road serves primarily as a through passage for road vehicles or (less frequently) pedestrians. Buskers, beggars, boulevardiers, patrons of pavement cafés, peoplewatchers, streetwalkers, and a diversity of other characters are habitual users of a street; the same people would not typically be found on a road.

In rural and suburban environments where street life is rare, the terms "street" and "road" are frequently considered interchangeable. Still, even here, what is called a "street" is usually a smaller thoroughfare, such as a road within a housing development feeding directly into individual driveways. In the last half of the 20th century these streets often abandoned the tradition of a rigid, rectangular grid, and instead were designed to discourage through traffic. This and other traffic calming methods provided quiet for families and play space for children. Adolescent suburbanites find, in attenuated form, the amenities of street life in shopping malls where vehicles are forbidden.

If a road connects places, then a street connects people. One may "hit the road" to see the wonders of the world—Jack Kerouac famously chronicled one such journey—but the latest bling will "hit the streets" before it ever appears on a road. It is "on the street" where one hears an interesting rumor, where one bumps into an old acquaintance, where one acquires street smarts. One seldom sees a "road" vendor except of fresh produce, or a "road" performer. You'll never find yourself on a long "street" to nowhere or under assault by a violent "road" gang, hence politicians seldom view with alarm the prevalence of "crime in the roads". The street, not the road is home to the homeless unless they are hoboes, and even Kerouac's hero finally returned to find his friends on a New York street.

A town square or plaza is a little more like a street, but a town square is rarely paved with asphalt and may not make any concessions for through traffic at all.

Nomenclature

Nevsky Prospekt, is the main street in the city of St. Petersburg, 1901.

An avenue in São Paulo.

There is a haphazard relationship, at best, between a thoroughfare's function and its name. For example, London's Abbey Road serves all the vital functions of a street, despite its name, and locals are more apt to refer to the "street" outside than the "road". A desolate road in rural Montana, on the other hand, may bear a sign proclaiming it "Davidson Street", but this does not make it a "street" except in the original sense of a paved road.

In the United Kingdom many towns will refer to their main thoroughfare as the High Street (in the United States it would be called the Main Street—however, occasionally "Main Street" in a city or town is a street other than the *de facto* main thoroughfare), and many of the ways leading off it will be named "Road" despite the urban setting. Thus the town's so-called "Roads" will actually be more street like than a road.

Some streets may even be called highways. For example, Hurontario Street in Mississauga, Ontario, Canada, is commonly referred to as "Highway 10"—even though such a highway designation no longer officially exists. This is probably because the street is a modern suburban arterial that was urbanized after decades of having the status and function a true highway, so people continued to use the number from force of habit.

In some other English-speaking countries, such as Australia and New Zealand, cities are often divided by a main "Road", with "Streets" leading from this "Road", or the cities are divided by thoroughfares known as "Streets" or "Roads" with no apparent differentiation between the two. In Auckland, for example, the main shopping precinct is located around Queen Street and Karangahape Road.

Streets have existed for as long as humans have lived in permanent settlements. However, modern civilization in much of the New World developed around transportation provided by motor vehicles. In some parts of the English-speaking world, such as North America, many think of the street as a thoroughfare for vehicular traffic first and foremost. In this view, pedestrian traffic is incidental to the street's purpose; a street consists of a thoroughfare running through the middle (in essence, a road), and may or may not have pavements (i.e., sidewalks) along the sides.

In an even narrower sense, some may think of a street as only the vehicle-driven and parking part of the thoroughfare. Thus, sidewalks (pavements) and road verges would not be thought of as part of the street. A mother may tell her toddlers, "Don't go out into the street, so you don't get hit by a car."

Among urban residents of the English-speaking world, the word "street" appears to carry its original connotations (i.e., the facilitation of traffic as a prime purpose, and "street life" as an incidental benefit). For instance, a *New York Times* writer lets casually slip the observation that automobile-laden Houston Street, in lower Manhattan, is "a street that can hardly be called 'street' anymore, transformed years ago into an eight-lane raceway that alternately resembles a Nascar event and a parking lot." Published in the paper's Metro section, the article evidently presumes an audience with an innate grasp of the modern urban role of the street. To the readers of the Metro section, vehicular traffic does not reinforce, but rather detracts from, the essential "street-ness" of a street.

At least one map has been made to illustrate the geography of naming conventions for thoroughfares; avenue, boulevard, circle, road, street, and other suffixes are compared and contrasted.

Highway

A highway is any public road or other public way on land. It is used for major roads, but also includes other public roads and public tracks: It is *not* an equivalent term to Controlled-access highway, or a translation for *autobahn*, *autoroute*, etc.

In North American and Australian English, major roads such as controlled-access highways or arterial roads are often state highways (Canada: provincial highways). Other roads may be designated "county highways" in the US and Ontario. These classifications refer to the level of government (state, provincial, county) that maintains the roadway.

In British English, "highway" is primarily a legal term. Everyday use normally implies roads, while the legal use covers any route or path with a public right of access, including footpaths etc.

The term has led to several related derived terms, including highway system, highway code, highway patrol and highwayman.

The term highway exists in distinction to "waterway".

Overview

Major highways are often named and numbered by the governments that typically develop and maintain them. Australia's Highway 1 is the longest national highway in the world at over 14,500 km or 9,000 mi and runs almost the entire way around the continent. China has the world's largest network of highways followed closely by the United States of America. Some highways, like the Pan-American Highway or the European routes, span multiple countries. Some major highway routes include ferry services, such as U.S. Route 10, which crosses Lake Michigan.

Traditionally highways were used by people on foot or on horses. Later they also accommodated carriages, bicycles and eventually motor cars, facilitated by advancements in road construction. In the 1920s and 1930s, many nations began investing heavily in progressively more modern highway systems to spur commerce and bolster national defense.

Major modern highways that connect cities in populous developed and developing countries usually incorporate features intended to enhance the road's capacity, efficiency, and safety to various degrees. Such features include a reduction in the number of locations for user access, the use of dual carriageways with two or more lanes on each carriageway, and grade-separated junctions with other roads and modes of transport. These features are typically present on highways built as *motorways* (*freeways*).

Terminology

England and Wales

The general legal definition deals with right of use not the form of construction; this is distinct from e.g. the popular use of the word in the US. A highway is defined in English common law by a number of similarly-worded definitions such as "a way over which all members of the public have the right to pass and repass without hindrance" usually accompanied by "at all times"; ownership of the ground is for most purposes irrelevant thus the term encompasses all such ways from the widest trunk roads in public ownership to the narrowest footpath providing unlimited pedestrian access over private land.

A highway might be open to all forms of lawful land traffic (i.e. vehicular, horse, pedestrian) or limited to specific types of traffic or combinations of types of traffic; usually a highway available to vehicles is available to foot or horse traffic, a highway available to horse traffic is available to pedestrians but exceptions can apply usually in the form of a highway only being available to vehicles or subdivided into dedicated parallel sections for different users.

A highway can share ground with a private right of way for which full use is not available to the general public as often will be the case with farm roads which the owner may use for any purpose but for which the general public only has a right of use on foot and/or horseback. The status of *highway* on most older roads has been gained by established public use while newer roads are typically *dedicated* as highways from the time they are adopted (taken into the care and control of a council or other public authority). In England and Wales, a public highway is also known as "*The Queen's Highway*".

The core definition of a highway is modified in various legislation for a number of purposes but only for the specific matters dealt with in each such piece of legislation. This is typically in the case of bridges, tunnels and other structures whose ownership, mode of use and/or availability would otherwise exclude them from the general definition of a highway, examples in recent years are commonly toll bridges and tunnels which have the definition of *highway* imposed upon them (in a legal order applying only to the individual structure) to allow application of most traffic laws to those using them but without causing all of the general obligations or rights of use otherwise applicable to a highway.

Scotland

Scots law is similar to English law with regard to highways but with differing terminology and legislation. What is defined in England as a *highway* will often in Scotland be what is defined by s.151 Roads (Scotland) Act 1984 (but only "in this act" although other legislation could imitate) simply as a road, that is :-

- "any way (other than a waterway) over which there is a public right of passage (by whatever means [and whether subject to a toll or not]) and includes the road's verge, and any bridge (whether permanent or temporary) over which, or tunnel through which, the road passes; and any reference to a road includes a part thereof; "

The word *highway* is itself no longer a statutory expression in Scots law but remains in common law.

United States

In American law, the word "highway" is sometimes used to denote any public way used for travel, whether a "road, street, and parkway"; however, in practical and useful meaning, a "highway" is a major and significant, well-constructed road that is capable of carrying reasonably heavy to extremely heavy traffic. Highways generally have a route number designated by the state and federal departments of transportation.

California Vehicle Code, Sections 360, 590, define a "highway" as only a way open for use of motor vehicles, but the California Supreme Court has held that "the definition of 'highway' in the Vehicle Code is used for special purposes of that act," and that canals of the Los Angeles neighborhood of Venice, California, are "highways" that are entitled to be maintained with state highway funds.

Smaller roads may be termed byways.

History

A German *Autobahn* in the 1930s.

Modern highway systems developed in the 20th century as the automobile gained popularity. The world's first limited access road was constructed on Long Island New York in the United States known as the Long Island Motor Parkway or the Vanderbilt Motor Parkway. It was completed in 1911. Construction of the Bonn–Cologne autobahn began in 1929 and was opened in 1932 by the mayor of Cologne, Konrad Adenauer.

In the USA, the Federal Aid Highway Act of 1921 (Phipps Act) enacted a fund to create an extensive highway system. In 1922, the first blueprint for a national highway system (the Pershing Map) was published. The Federal Aid Highway Act of 1956 allocated $25 billion for the construction of the 41,000 miles (66 Mm) long Interstate Highway System over a 20-year period.

In Great Britain, the Special Roads Act 1949 provided the legislative basis for roads for restricted classes of vehicles and non-standard or no speed limits applied (later mostly termed motorways but now with speed limits not exceeding 70 mph); in terms of general road law this legislation overturned the usual principle that a road available to vehicular traffic was also available to horse or pedestrian traffic as is usually the only practical change when non-motorways are reclassified as *special roads*. The first section of motorway in the UK opened in 1958 (part of the M6 motorway) and then in 1959 the first section of the M1 motorway.

Social Effects

Reducing travel times relative to city or town streets, modern highways with limited access and grade separation create increased opportunities for people to travel for business, trade or pleasure and also provide trade routes for goods. Modern highways reduce commute and other travel time but additional road capacity can also release latent traffic demand. If not accurately predicted at the planning stage, this extra traffic may lead to the new road becoming congested sooner than would otherwise be anticipated by considering increases in vehicle ownership. More roads allow drivers to use their cars when otherwise alternatives may have been sought, or the journey may not have been made, which can mean that a new road brings only short-term mitigation of traffic congestion.

Where highways are created through existing communities, there can be reduced community cohesion and more difficult local access. Consequently, property values have decreased in many cut-off neighborhoods, leading to decreased housing quality over time.

Economic Effects

In transport, demand can be measured in numbers of journeys made or in total distance travelled across all journeys (e.g. passenger-kilometres for public transport or vehicle-kilometres of travel (VKT) for private transport). Supply is considered to be a measure of capacity. The price of the good (travel) is measured using the generalised cost of travel, which includes both money and time expenditure.

The effect of increases in supply (capacity) are of particular interest in transport economics, as the potential environmental consequences are significant.

In addition to providing benefits to their users, transport networks impose both positive and negative externalities on non-users. The consideration of these externalities—particularly the negative ones—is a part of transport economics. Positive externalities of transport networks may include the ability to provide emergency services, increases in land value and agglomeration benefits. Negative externalities are wide-ranging and may include local air pollution, noise pollution, light pollution, safety hazards, community severance and congestion. The contribution of transport systems to potentially hazardous climate change is a significant negative externality which is difficult to evaluate quantitatively, making it difficult (but not impossible) to include in transport economics-based research and analysis. Congestion is considered a negative externality by economists.

A 2016 study finds that for the United States "a 10% increase in a region's stock of highways causes a 1.7% increase in regional patenting over a five-year period."

Environmental Effects

Noise, light and air pollution are negative environmental effects highways can have on their surroundings.

Highways are extended linear sources of pollution.

Roadway noise increases with operating speed so major highways generate more noise than arterial streets. Therefore, considerable noise health effects are expected from highway systems. Noise mitigation strategies exist to reduce sound levels at nearby sensitive receptors. The idea that highway design could be influenced by acoustical engineering considerations first arose about 1973.

Air quality issues: Highways may contribute fewer emissions than arterials carrying the same vehicle volumes. This is because high, constant-speed operation creates an emissions reduction compared to vehicular flows with stops and starts. However, concentrations of air pollutants near highways may be higher due to increased traffic volumes. Therefore, the risk of exposure to elevated levels of air pollutants from a highway may be considerable, and further magnified when highways have traffic congestion.

New highways can also cause habitat fragmentation, encourage urban sprawl and allow human intrusion into previously untouched areas, as well as (counterintuitively) increasing congestion, by increasing the number of intersections.

They can also reduce the use of public transport, indirectly leading to greater pollution.

High-occupancy vehicle lanes are being added to some newer/reconstructed highways in North America and other countries around the world to encourage carpooling and mass-transit. These lanes help reduce the number of cars on the highway and thus reduces pollution and traffic congestion by promoting the use of carpooling in order to be able to use these lanes. However, they tend to require dedicated lanes on a highway, which makes them difficult to construct in dense urban areas where they are the most effective.

To address habitat fragmentation, wildlife crossings have become increasingly popular in many countries. Wildlife crossings allow animals to safely cross human-made barriers like highways.

Road Traffic Safety

Road traffic safety describes the safety performance of roads and streets, and methods used to reduce the harm (deaths, injuries, and property damage) on the highway system from traffic collisions. It includes the design, construction and regulation of the roads, the vehicles used on them and the training of drivers and other road-users.

A report published by the World Health Organization in 2004 estimated that some 1.2m people were killed and 50m injured on the roads around the world each year and was the leading cause of death among children 10–19 years of age.

The report also noted that the problem was most severe in developing countries and that simple prevention measures could halve the number of deaths. For reasons of clear data collection, only harm involving a road vehicle is included. A person tripping with fatal consequences or dying for some unrelated reason on a public road is not included in the relevant statistics.

Statistics

International sign used widely in Europe denoting the start of special restrictions for a section of highway.

The United States has the world's largest network of highways, including both the Interstate Highway System and the U.S. Highway System. At least one of these networks is present in every state and they interconnect most major cities.

China's highway network is the second most extensive in the world, with a total length of about 3.573 Gm. China's expressway network is the longest Expressway system in the world, and it is quickly expanding, stretching some 85 Mm at the end of 2011. In 2008 alone, 6.433 Mm expressways were added to the network.

The Cross Bronx Expressway in New York, United States uses asphalt and concrete pavement, both of which are popular road surfaces on highways.

Longest International Highway

The Pan-American Highway, which connects many countries in the Americas, is nearly 25,000 kilometres (15,534 mi) long as of 2005. The Pan-American Highway is discontinuous because there is a significant gap in it in southeastern Panama, where the rainfall is immense and the terrain is entirely unsuitable for highway construction.

Longest National Highway (Point to Point)

Trans-Canada Highway has two routes, with the northern Route spanning 7,821 km (4,860 mi) long as of 2006 alone, and over 10,700 km long including the southern portion. The T.C.H. runs east-west across southern Canada, the populated portion of the country, and it connects many of the major urban centres along its route crossing almost all of the provinces, and reaching almost all of the capital cities. The T.C.H. begins on the east coast in Newfoundland, traverses that island, and crosses to the mainland by ferry. It reaches most of the Maritime Provinces of eastern Canada, and a side route using ferries traverses the province of Prince Edward Island. After crossing the two most populous provinces of Quebec and Ontario, the T.C.H. continues westward across Manitoba, Saskatchewan, Alberta, and British Columbia. After reaching Vancouver, B.C., on the Pacific Coast, there is a ferry route west to Vancouver Island and the provincial capital city of Victoria, B.C.

Longest National Highway (Circuit)

Australia's Highway 1 at over 20 Mm (12,000 mi). It runs almost the entire way around the continent's coastline. With the exception of the Federal Capital of Canberra, which is far inland, Highway 1 links all of Australia's capital cities, although Brisbane and Darwin are not directly connected, but rather are bypassed short distances away. Also, there is a ferry connection to the island state of Tasmania, and then a stretch of Highway 1 that links the major towns and cities of Tasmania, including Launceston and Hobart (this state's capital city).

Largest National Highway System

The United States of America has approximately 6.43 gigametres (3,995,417 mi) of highway within its borders as of 2008.

Busiest Highway

Highway 401 in Ontario, Canada, has volumes surpassing an average of 500,000 vehicles per day in some sections of Toronto as of 2006.

Widest Highway (Maximum Number of Lanes)

The Katy Freeway (part of Interstate 10) in Houston, Texas, has a total of 26 lanes in some sections as of 2007. However, they are divided up into general use/ frontage roads/ HOV lanes, restricting the traverse traffic flow.

Widest Highway (Maximum Number of Through Lanes)

Interstate 5 along a two-mile-long (3.2 km) section between Interstate 805 and California State Route 56 in San Diego, California, which was completed in April 2007, is 22 lanes wide.

Highest International Highway

The Karakoram Highway, between Pakistan and China, is at an altitude of 4,693 metres (15,397 ft).

Bus lane

Highway bus lane on Gyeongbu Expressway in South Korea.

Some countries incorporate bus lanes onto highways.

Country	Highway	Bus lanes (km)	Section
Australia	M2 Hills Motorway		Abbott Road–Beecroft Road (Sydney)
India	National Highway (India)	19	30 lanes Road (Mumbai)
Canada	Don Valley Parkway	0.458	shoulder converted as bypass lane from Lawrence Avenue East to York Mills Road
Canada	Ontario Highway 417	7	Eagleson Road–Ontario Highway 417 (Ottawa)
Canada	Ontario Highway 403	6	Mavis Road–Winston Churchill Boulevard (Mississauga)
Hong Kong	Tuen Mun Road		
South Korea	Gyeongbu Expressway	137.4	Hannam IC (Seoul) ~ Sintanjin IC (Daejeon)
Netherlands	A1 motorway (Netherlands)		End of A6-Vechtbrug (Muiden)

South Korea

In South Korea, in February 1995—Bus lane (essentially an HOV-9) established between the northern terminus and Sintanjin for important holidays and on 1 July 2008—Bus lane enforcement between Seoul and Osan (Sintanjin on weekends) becomes daily between 6 AM and 10 PM. On 1 October this is adjusted to 7 AM to 9 PM weekdays, 9 AM to 9 PM weekends.

India

World Secondly Highway in India After in USA Indian Highway Is Number One in Worldwide Indian biggest highway Number NH 7 (Mumbai-New Delhi) his 70 Lean in World biggest

Hong Kong

In Hong Kong, some highways are set up with bus lanes to solve the traffic congestion.

District	Highway	Section
Tuen Mun	9 Tuen Mun Road	So Kwun Wat to Sham Tseng
Sha Tin	1 Lion Rock Tunnel	The entry of the tunnel

Pakistan

The transport sector accounts for 10.5% of Pakistan's GDP, but its road infrastructure is worse than that of India, Bangladesh and Indonesia.

Philippines

Traffic congestion was a principal problem in major roads and highways in the Philippines, especially in Metro Manila and other major cities. The government decided to set up some bus lanes in Metro Manila like in the Epifanio delos Santos Avenue.

Bike Lane

Bike lanes (US) or cycle lanes (UK) are types of bikeways (cycleways) with lanes on the roadway for cyclists only. In the United Kingdom, an on-road cycle-lane can be restricted [to cycles] (marked with a solid white line, entry by motor vehicles is prohibited) or advisory (marked with a broken white line, entry by motor vehicles is permitted). In the United States, a *designated bicycle lane* (1988 MUTCD) or *class II bikeway* (Caltrans) is always marked by a solid white stripe on the pavement and is for 'preferential use' by bicyclists. There is also a *class III bicycle route*, which has roadside signs suggesting a route for cyclists, and urging sharing the road.

In France, segregated cycling facilities on the carriageway are called *bande cyclable*, those beside the carriageway or totally independent ones *piste cyclable*, all together *voie cyclable*. In Belgium, traffic laws do not distinguish cycle lanes from cyclepaths. Cycle lanes are marked by *two* parallel broken white lines, and they are defined as being "not wide enough to allow use by motor vehicles". Note that there is some confusion possible here: both in French (*piste cyclable*) and in Dutch (*fietspad*) the term for these lanes can also denote a segregated cycle track, marked by a road sign; the cycle lane is therefore often referred to as a "piste cyclable marquée" (in French) or a "gemarkeerd fietspad" (in Dutch), i.e. a cycle lane/track which is "marked" (i.e. identified by road markings) rather than one which is identified by a road sign. In the Netherlands the cycle lane is normally called "fietsstrook" instead of "fietspad".

Cycle lane can be seen as part of the road on both sides separated by a solid white line (Moncton, New Brunswick, Canada).

Cycle lane in Počernická street in Prague 10, Czech Republic

Public Transport

Public transport (also known as public transportation, public transit, or mass transit) is a shared passenger-transport service which is available for use by the general public, as distinct from modes such as taxicab, carpooling, or hired buses, which are not shared by strangers without private arrangement.

Public transport modes include city buses, trolleybuses, trams (or light rail) and passenger trains, rapid transit (metro/subways/undergrounds etc) and ferries. Public transport between cities is dominated by airlines, coaches, and intercity rail. High-speed rail networks are being developed in many parts of the world. Most public transport runs to a scheduled timetable with the most frequent services running to a headway (e.g.: "every 5 minutes" as opposed to being scheduled for any specific time of the day). Share taxis offer on-demand services in many parts of the world, and some services will wait until the vehicle is full before it starts. Paratransit is sometimes used in areas of low-demand and for people who need a door-to-door service.

There are distinct differences in urban public transit between Asia, North America, and Europe. In Asia, mass transit operations are predominantly run by profit-driven privately owned and publicly traded mass transit and real estate conglomerates. In North America, mass transit operations are predominantly run by municipal transit authorities. In Europe, mass transit operations are predominantly run by both state-owned and private companies. Public transport services can be profit-driven by use of pay-by-the-distance fares or funded by government subsidies in which flat rate fares are charged to each passenger. Services can be fully profitable through high ridership numbers and high farebox recovery ratios, or can be regulated and possibly subsidized from local or national tax revenue. Fully subsidized, zero-fare (free) services operate in some towns and cities.

For historical and economic reasons, there are differences internationally regarding use and extent of public transport. While countries in the Old World tend to have extensive and frequent systems serving their old and dense cities, many cities of the New World have more sprawl and much less comprehensive public transport. The International Association of Public Transport (UITP) is the international network for public transport authorities and operators, policy decision-makers, scientific institutes and the public transport supply and service industry. It has 3,400 members from 92 countries.

History

Conveyances designed for public hire are as old as the first ferries, and the earliest public transport was water transport: on land people walked (sometimes in groups and on pilgrimages, as noted in sources such as the Bible and *The Canterbury Tales*) or (at least in the Old World) rode an animal. Ferries appear in Greek mythology—corpses in ancient Greece were buried with a coin underneath their tongue to pay the ferryman Charon to take them to Hades.

Some historical forms of public transport include the stagecoach, traveling a fixed route from coaching inn to coaching inn, and the horse-drawn boat carrying paying passengers, which was a feature of European canals from their 17th-century origins. (The canal itself as a form of infrastructure dates back to antiquity – ancient Egyptians certainly used a canal for freight transportation to bypass the Aswan cataract – and the Chinese also built canals for water transportation as far back as the Warring States period which began in the 5th century BCE. Whether or not those canals were used for for-hire public transport remains unknown; the Grand Canal in China (begun in 486 BCE) served primarily for shipping grain.)

Early trolley car in Newton, Massachusetts

The omnibus, the first organized public transit system within a city, appears to have originated in Paris, France, in 1662, although the service in question failed a few months after its founder, Blaise Pascal, died in August 1662; omnibuses are next known to have appeared in Nantes, France, in 1826. The omnibus was introduced to London in July 1829.

The first passenger horse-drawn railway opened in 1806: it ran between Swansea and Mumbles in southwest Wales in the United Kingdom. In 1825 George Stephenson built the *Locomotion* for the Stockton and Darlington Railway in northeast England, the first public steam railway in the world.

Types

Airline

An airline provides scheduled service with aircraft between airports. Air travel has high up to very high speeds, but incurs large waiting times prior and after travel, and is therefore often only feasible over longer distances or in areas where lack of ground infrastructure makes other modes of transport impossible. Bush airlines work more similar to bus stops; an aircraft waits for passengers and takes off when the aircraft is full of luggage.

Bus and Coach

Bus services use buses on conventional roads to carry numerous passengers on shorter journeys. Buses operate with low capacity (i.e. compared with trams or trains), and can operate on conventional roads, with relatively inexpensive bus stops to serve passengers. Therefore buses are commonly used in smaller cities, towns, and rural areas, as well as for shuttle services supplementing other means of transit in large cities. Bus rapid transit is an ambiguous term used for buses operating on dedicated right-of-way, much like a light rail. Trolleybuses are electric buses that employ overhead wires to get power for traction. Online Electric Vehicles are buses that run on a conventional battery, but are recharged frequently at certain points via underground wires.

TransMilenio buses in Bogotá, Colombia

Masivo Integrado de Occidente (MIO) in Cali, Colombia

Trolza trolleybus in Moscow – operating the world's largest trolleybus system

Coach services use coaches (long-distance buses) for suburb-to-CBD or longer-distance transportation. The vehicles are normally equipped with more comfortable seating, a separate luggage compartment, video and possibly also a toilet. They have higher standards than city buses, but a limited stopping pattern.

Train

Passenger rail transport is the conveyance of passengers by means of wheeled vehicles specially designed to run on railways. Trains allow high capacity on short or long distance, but require track, signalling, infrastructure and stations to be built and maintained. Urban rail transit consists of trams, light rail, rapid transit, people movers, commuter rail, monorail suspension railways and funiculars.

3 different types of Trains (U-Bahn, S-Bahn and Regional/Longdistance trains) in Hamburg, Germany

Left: Tokyo's Yamanote Line, one of the world's busiest commuter rail lines. Right: Chicago Transit Authority control tower 18 guides elevated Chicago 'L' northbound Purple and Brown lines intersecting with westbound Pink and Green lines and the looping Orange line above the Wells and Lake street intersection in the loop.

Commuter, Intercity, and High-speed Rail

A SEPTA Regional Rail train in Cheltenham, Pennsylvania, a form of commuter rail.

Commuter rail is part of an urban area's public transport; it provides faster services to outer suburbs and neighboring towns and villages. Trains stop at stations that are located to serve a smaller suburban or town center. The stations are often combined with shuttle bus or park and ride systems. Frequency may be up to several times per hour, and commuter rail systems may either be part of the national railway or operated by local transit agencies.

Intercity rail is long-haul passenger services that connect multiple urban areas. They have few stops, and aim at high average speeds, typically only making one of a few stops per city. These services may also be international.

High-speed rail is passenger trains operating significantly faster than conventional rail—typically defined as at least 200 kilometres per hour (120 mph). The most predominant systems have been built in Europe and Japan, and compared with air travel, offer long-distance rail journeys as quick as air services, have lower prices to compete more effectively and use electricity instead of combustion.

Tram and Light Rail

Tram of Toronto, which operates the largest streetcar system in North America

Trams are railborne vehicles that run in city streets or dedicated tracks. They have higher capacity than buses, but must follow dedicated infrastructure with rails and wires either above or below the track, limiting their flexibility.

Light rail is a modern development (and use) of the tram, with dedicated right-of-way not shared with other traffic, (often) step-free access and increased speed. Light rail lines are, thus, essentially modernized interurbans.

Rapid Transit

SkyTrain automated rapid transit system in Vancouver, Canada

A rapid transit railway system (also called a metro, underground, or subway) operates in an urban area with high capacity and frequency, and grade separation from other traffic.

Systems are able to transport large amounts of people quickly over short distances with little land use. Variations of rapid transit include people movers, small-scale light metro and the commuter rail hybrid S-Bahn. More than 160 cities have rapid transit systems, totalling more than 8,000 km (4,971 mi) of track and 7,000 stations. Twenty-five cities have systems under construction.

Personal Rapid Transit

people mover vehicle of Morgantown Personal Rapid Transit

Personal rapid transit is an automated cab service that runs on rails or a guideway. This is an uncommon mode of transportation (excluding elevators) due to the complexity of automation. A fully implemented system might provide most of the convenience of individual automobiles with the efficiency of public transit. The crucial innovation is that the automated vehicles carry just a few passengers, turn off the guideway to pick up passengers (permitting other PRT vehicles to continue at full speed), and drop them off to the location of their choice (rather than at a stop). Conventional transit simulations show that PRT might attract many auto users in problematic medium-density urban areas. A number of experimental systems are in progress. One might compare personal rapid transit to the more labor-intensive taxi or paratransit modes of transportation, or to the (by now automated) elevators common in many publicly accessible areas.

Cable-propelled Transit

Cable-propelled transit (CPT) is a transit technology that moves people in motor-less, engine-less vehicles that are propelled by a steel cable. There are two sub-groups of CPT – gondola lifts and cable cars (railway). Gondola lifts are supported and propelled from above by cables, whereas cable cars are supported and propelled from below by cables.

While historically associated with usage in ski resorts, gondola lifts are now finding increased consumption and utilization in many urban areas – built specifically for the purposes of mass transit. Many, if not all, of these systems are implemented and fully integrated within existing public transportation networks. Examples include Metrocable (Medellín), Metrocable (Caracas), Portland Aerial Tram, Roosevelt Island Tramway in New York City, and London's Emirates Air Line.

Metrocable gondola lifts of Medellín, Colombia

Ferry

water taxi (*vaporetto*) of Venice, Italy

A ferry is a boat or ship, used to carry (or *ferry*) passengers, and sometimes their vehicles, across a body of water. A foot-passenger ferry with many stops is sometimes called a water bus. Ferries form a part of the public transport systems of many waterside cities and islands, allowing direct transit between points at a capital cost much lower than bridges or tunnels, though at a lower speed. Ship connections of much larger distances (such as over long distances in water bodies like the Mediterranean Sea) may also be called ferry services.

Auto Rickshaws

Auto rickshaws carry people and goods in many developing countries. Also known as a three-wheeler, Samosa, tempo, tuk-tuk, trishaw, auto, rickshaw, autorick, bajaj, rick, tricycle, mototaxi, baby taxi or lapa in popular parlance, they are motorized version of the traditional pulled rickshaw or cycle rickshaw. They are an essential form of urban transport, both as vehicles for hire and for private use, in many developing countries, and a form of novelty transport in many Eastern countries.

Auto rickshaw in Lahore, Pakistan

Operation

Timelapse video of Downtown Seattle from atop a Community Transit double-decker bus

Infrastructure

All public transport runs on infrastructure, either on roads, rail, airways or seaways. The infrastructure can be shared with other modes, freight and private transport, or it can be dedicated to public transport. The latter is especially valuable in cases where there are capacity problems for private transport. Investments in infrastructure are expensive and make up a substantial part of the total costs in systems that are new or expanding. Once built, the infrastructure will require operating and maintenance costs, adding to the total cost of public transport. Sometimes governments subsidize infrastructure by providing it free of charge, just as is common with roads for automobiles.

Interchanges

Interchanges are locations where passengers can switch from one public transport route to another. This may be between vehicles of the same mode (like a bus interchange), or e.g. between bus and train. It can be between local and intercity transport (such as at a central station or airport).

Timetables

Timetables (or 'schedules' in North American English) are provided by the transport operator to allow users to plan their journeys. They are often supplemented by maps and fare schemes to help travelers coordinate their travel. Online public transport route planners help make planning easier. Mobile apps are available for multiple transit systems that provide timetables and other service information and, in some cases, allow ticket purchase, some allowing to plan your journey, with time fares zones eg.

Services are often arranged to operate at regular intervals throughout the day or part of the day (known as clock-face scheduling). Often, more frequent services or even extra routes are operated during the morning and evening rush hours. Coordination between services at interchange points is important to reduce the total travel time for passengers. This can be done by coordinating shuttle services with main routes, or by creating a fixed time (for instance twice per hour) when all bus and rail routes meet at a station and exchange passengers. There is often a potential conflict between this objective and optimising the utilisation of vehicles and drivers.

Financing

The main sources of financing are ticket revenue, government subsidies and advertising. The percentage of revenue from passenger charges is known as the farebox recovery ratio. A limited amount of income may come from land development and rental income from stores and vendors, parking fees, and leasing tunnels and rights-of-way to carry fiber optic communication lines.

Fare and Ticketing

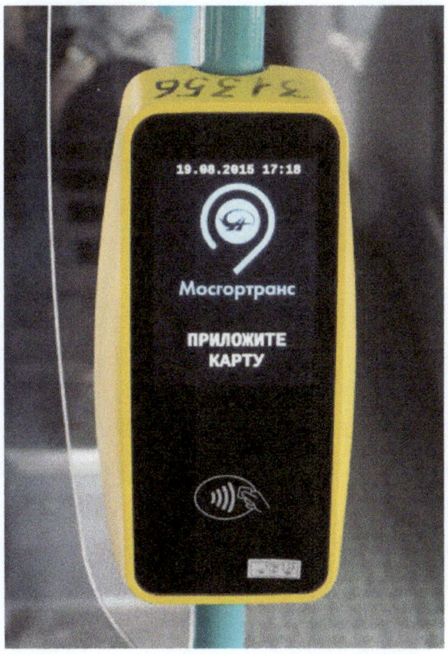

A contactless ticket validator used in Moscow, Russia

Most—but not all—public transport requires the purchase of a ticket to generate revenue for the operators. Tickets may be bought either in advance, or at the time of the journey, or the carrier may allow both methods. Passengers may be issued with a paper ticket, a metal or plastic token, or a magnetic or electronic card (smart card, contactless smart card). Sometimes a ticket has to be validated, e.g. a paper ticket has to be stamped, or an electronic ticket has to be checked in.

Tickets may be valid for a single (or return) trip, or valid within a certain area for a period of time. The fare is based on the travel class, either depending on the traveled distance, or based on zone pricing.

The Access card is a smart card for public transportation tickets in Stockholm, Sweden.

The tickets may have to be shown or checked automatically at the station platform or when boarding, or during the ride by a conductor. Operators may choose to control all riders, allowing sale of the ticket at the time of ride. Alternatively, a proof-of-payment system allows riders to enter the vehicles without showing the ticket, but riders may or may not be controlled by a ticket controller; if the rider fails to show proof of payment, the operator may fine the rider at the magnitude of the fare.

Multi-use tickets allow travel more than once. In addition to return tickets, this includes period cards allowing travel within a certain area (for instance month cards), or during a given number of days that can be chosen within a longer period of time (for instance eight days within a month). Passes aimed at tourists, allowing free or discounted entry at many tourist attractions, typically include zero-fare public transport within the city. Period tickets may be for a particular route (in both directions), or for a whole network. A free travel pass allowing free and unlimited travel within a system is sometimes granted to particular social sectors, for example students, elderly, children, employees (*job ticket*) and the physically or mentally disabled.

Zero-fare public transport services are funded in full by means other than collecting a fare from passengers, normally through heavy subsidy or commercial sponsorship by businesses. Several mid-size European cities and many smaller towns around the world have converted their entire bus networks to zero-fare. The only European capital with free public transport is Tallinn. Local zero-fare shuttles or inner-city loops are far more common than city-wide systems. There are also zero-fare airport circulators and university transportation systems.

Revenue, Profit and Subsidies

Governments frequently opt to subsidize public transport, for social, environmental or economic reasons. Common motivations include the desire to provide transport to people who are unable to use an automobile, and to reduce congestion, land use and automobile emissions. Other motives may include promoting business and economic growth, or urban renewal in formerly deprived areas of the city. Public transit systems rarely operate without government support. Some systems are owned and operated by a government agency; other transportation services may be commercial, but receive special benefits from the government compared to a normal company.

Subsidies may take the form of direct payments for financially unprofitable services, but support may also include indirect subsidies. For example, the government may allow free or reduced-cost use of state-owned infrastructure such as railways and roads, to stimulate public transport's economic competitiveness over private transport, that normally also has free infrastructure (subsidized through such things as gas taxes). Other subsidies include tax advantages (for instance aviation fuel is typically not taxed), bailouts if companies that are likely to collapse (often applied to airlines) and reduction of competition through licensing schemes (often applied to taxis and airlines). Private transport is normally subsidized indirectly through free roads and infrastructure, as well as incentives to build car factories and, on occasion, directly via bailouts of automakers.

Land development schemes may be initialized, where operators are given the rights to use lands near stations, depots, or tracks for property development. For instance, in Hong Kong, MTR Corporation Limited and KCR Corporation generate additional profits from land development to partially cover the cost of the construction of the urban rail system.

Some supporters of mass transit believe that use of taxpayer capital to fund mass transit will ultimately save taxpayer money in other ways, and therefore, state-funded mass transit is a benefit to the taxpayer. Some research has supported this position, but the measurement of benefits and costs is a complex and controversial issue. A lack of mass transit results in more traffic, pollution, and road construction to accommodate more vehicles, all costly to taxpayers; providing mass transit will therefore alleviate these costs. (Perhaps, although right-wing think tanks disagree)

Safety and Security

A LASD deputy and a police dog patrol a LA Metro light rail train.

In the United States expansion of public transportation systems is often opposed by critics who see them as vehicles for violent criminals and homeless persons to expand into new areas (to which they would otherwise have to walk). According to the Transportation Research Board, "[v]iolent crime is perceived as pandemic .... Personal security affects many peoples' [sic] decisions to use public transportation." Despite the occasional highly publicized incident, the vast majority of modern public transport systems are well designed and patrolled and generally have low crime rates. Many systems are monitored by CCTV, mirrors, or patrol.

Nevertheless, some systems attract vagrants who use the stations or trains as sleeping shelters, though most operators have practices that discourage this.

Though public transit accidents attract far more publicity than car wrecks, public transport has much lower accident rates. Annually, public transit prevents 200,000 deaths, injuries, and accidents had equivalent trips been made by car. The National Safety Council estimates riding the bus as over 170 times safer than private car.

Impact

Environmental

Although there is continuing debate as to the true efficiency of different modes of transportation, mass transit is generally regarded as significantly more energy efficient than other forms of travel. A 2002 study by the Brookings Institution and the American Enterprise Institute found that public transportation in the U.S uses approximately half the fuel required by cars, SUV's and light trucks. In addition, the study noted that "private vehicles emit about 95 percent more carbon monoxide, 92 percent more volatile organic compounds and about twice as much carbon dioxide and nitrogen oxide than public vehicles for every passenger mile traveled".

The pink Bus Rapid Transit of Metz uses a diesel-electric hybrid driving system, developed by Belgian Van Hool manufacturer.

A 2004 study from Lancaster University concluded that there was no environmental benefit to be gained from persuading car or plane travelers to switch to trains. Environmental group Friends of the Earth were skeptical of the findings, claiming the results are not in line with previous studies. The study showed that trains had failed to keep up with the advances that the automotive and aviation industries had made in improved fuel efficiency. Express trains travelling from London to Edinburgh consumed 11.5 litres more fuel per seat than a modern diesel car. A representative from *Modern Railways* magazine is reported as having said:

"I know this will generate howls of protest, but at present a family of four going by car is about as environmentally friendly as you can get."

Studies have shown that there is a strong inverse correlation between urban population density and energy consumption per capita, and that public transport could facilitate increased urban population densities, and thus reduce travel distances and fossil fuel consumption.

Supporters of the green movement usually advocate public transportation, because it offers decreased airborne pollution compared to automobiles. A study conducted in Milan, Italy in 2004 during and after a transportation strike serves to illustrate the impact that mass transportation has on the environment. Air samples were taken between 2 and 9 January, and then tested for Methane, Carbon Monoxide, non-methane Hydrocarbons (NMHCs), and other gases identified as harmful to the environment. The figure below is a computer simulation showing the results of the study "with 2 January showing the lowest concentrations as a result of decreased activity in the city during the holiday season. 9 January showed the highest NMHC concentrations because of increased vehicular activity in the city due to a public transportation strike."

Based on the benefits of public transport, the green movement has impacted public policy. For example, the state of New Jersey released *Getting to Work: Reconnecting Jobs with Transit*. This initiative attempts to relocate new jobs into areas with higher public transportation accessibility. The initiative cites the use of public transportation as being a means of reducing traffic congestion, providing an economic boost to the areas of job relocation, and most importantly, contributing to a green environment by reducing carbon dioxide (CO2) emissions.

Using public transportation can result in a reduction of an individual's carbon footprint. A single person, 20-mile (32 km) round trip by car can be replaced using public transportation and result in a net CO2 emissions reduction of 4,800 pounds (2,200 kg) per year. Using public transportation saves CO2 emissions in more ways than simply travel as public transportation can help to alleviate traffic congestion as well as promote more efficient land use. When all three of these are considered, it is estimated that 37 million metric tons of CO2 will be saved annually. Another study claims that using public transit instead of private in the U.S. in 2005 would have reduced CO2 emissions by 3.9 million metric tons and that the resulting traffic congestion reduction accounts for an additional 3.0 million metric tons of CO2 saved. This is a total savings of about 6.9 million metric tons per year given the 2005 values.

In order to compare energy impact of public transportation to private transportation, the amount of energy per passenger mile must be calculated. The reason that comparing the energy expenditure per person is necessary is to normalize the data for easy comparison. Here, the units are in per 100 p-km (read as person kilometer or passenger kilometer). In terms of energy consumption, public transportation is better than individual transport in a personal vehicle. In England, bus and rail are popular methods of public transportation, especially London. Rail provides rapid movement into and out of the city of London while busing helps to provide transport within the city itself. As of 2006–2007, the total energy cost of London's trains was 15 kWh per 100 p-km, about 5 times better than a personal car. For busing in London, it was 32 kWh per 100 p-km, or about 2.5 times that of a personal car. This includes lighting, depots, inefficiencies due to capacity (i.e., the train or bus may not be operating at full capacity at all times), and other inefficiencies. Efficiencies of transport in Japan in 1999 were 68 kWh per 100 p-km for a personal car, 19 kWh per 100 p-km for a bus, 6 kWh per 100 p-km for rail, 51 kWh per 100 p-km for air, and 57 kWh per 100 p-km for sea. These numbers from either country can be used in energy comparison calculations and/or life cycle assessment calculations.

Public transportation also provides an arena to test environmentally friendly fuel alternatives, such as hydrogen-powered vehicles. Swapping out materials to create lighter public transportation vehicles with the same or better performance will increase environmental friendliness of public transportation vehicles while maintaining current standards or improving them. Informing the public about the positive environmental effects of using public transportation in addition to pointing out the potential economic benefit is an important first step towards making a difference.

Land use

Urban space is a precious commodity and public transport utilises it more efficiently than a car dominant society, allowing cities to be built more compactly than if they were dependent on automobile transport. If public transport planning is at the core of urban planning, it will also force cities to be built more compactly to create efficient feeds into the stations and stops of transport. This will at the same time allow the creation of centers around the hubs, serving passengers' daily commercial needs and public services. This approach significantly reduces urban sprawl.

Traffic jam in São Paulo, Brazil

Social

A developed country is not a place where the poor have cars; it's where the rich use public transport —Enrique Penalosa, former mayor of Bogotá

An important social role played by public transport is to ensure that all members of society are able to travel, not just those with a driving license and access to an automobile—which include groups such as the young, the old, the poor, those with medical conditions, and people banned from driving. Automobile dependency is a name given by policy makers to places where those without access to a private vehicle do not have access to independent mobility.

Above that, public transportation opens to its users the possibility of meeting other people, as no concentration is diverted from interacting with fellow-travelers due to any steering activities.

Adding to the above-said, public transport becomes a location of inter-social encounters across all boundaries of social, ethnic and other types of affiliation.

Economic

Tram tunnel in Bratislava (Slovakia) - a connection of embankment with a city centre.

Public transport allows transport at an economy of scale not available through private transport. Advocates of public transport claim that investing in mass transit will ultimately reduce the total transport cost for the public. Time saved can also be significant, as less cars can translate to less congestion, and faster speeds for remaining motorists. Transit-oriented development can both improve the usefulness and efficiency of the public transit system as well as result in increased business for commercial developments.

Because of the increased traffic and access to transit systems, putting in public transit frequently has a positive effect on real estate prices. For example, the Washington DC Metro system has increased land desirability around its stations, and The Hong Kong metro MTR generates a profit by redeveloping land around and above its stations. Much public opposition to new transit construction can be based on the concern about the impact on neighborhoods of this new economic development. Few localities have the ability to seize and reassign development rights to a private transit operator, as Hong Kong has done.

Investment in public transport also has secondary positive effects on the local economy, with between $4 and $9 of economic activity resulting from every dollar spent. Many businesses rely on access to a transit system, in particular in cities and countries where access to cars is less widespread, businesses which require large amounts of people going to a same place may not be able to accommodate a large number of cars (concert venues, sport stadia, airports, exhibitions centres,...), or businesses where people are not able to use a car (bars, hospitals, or industries in the tourism sector whose customers may not have their cars). Transit systems also have an effect on derived businesses: commercial websites have been founded, such as Hopstop.com, that give directions through mass transit systems; in some cities, such as London, products themed on the local transport system are a popular tourist souvenir. Research in the Washington, DC area shows that public transport does a better job of providing high-skill residents with access to high-skill jobs than it does mid-skill residents to mid-skill jobs and low-skill residents to low-skill jobs.

However, public transport projects frequently have very large upfront costs, requiring large investments from either local government or private investors. Initial estimates of project cost and timescale are frequently underestimated, and nearly all public transport requires government subsidies and/or direct government support in order to remain operational.

The existence of a transit system can lower land values in some cases, either through influence on a region's demographics and crime rate (actual or perceived), or simply through the ambient noise and other discomforts the system creates.

Regulations

Food and Drink

Longer distance public transport sometimes sell food and drink on board, and/or have a dedicated buffet car and/or dining car. However, some urban transport systems forbid the consumption of food, drink, or even chewing gum when riding on public transport. Sometimes only types of food are forbidden with more risk of making the vehicles dirty, e.g. ice creams and French fries, and sometimes potato chips.

Some systems prohibit carrying open food or beverage containers, even if the food or beverage is not being consumed during the ride.

Smoking

In Australia, Canada, India, New Zealand, Norway, Switzerland, the United States, and most of the European Union, smoking is prohibited in all or some parts of most public transportation systems due to safety and health issues. Generally smoking is not allowed on buses and trains, while rules concerning stations and waiting platforms differ from system to system. The situation in other countries varies widely.

Noise

Many mass transit systems prohibit the use of audio devices, such as radios, CD players, and MP3 players unless used with an earphone through which only the user can hear the device.

Some mass transit systems have restricted the use of mobile phones. Long distance train services, such as the Amtrak system in the US, have "quiet cars" where mobile phone usage is prohibited.

Some systems prohibit passengers from engaging in conversation with the operator. Others require that passengers who engage in any conversation must keep the noise level low enough that it not be audible to other passengers.

Some systems have regulations on the use of profanity. In the United States, this has been challenged as a free speech issue.

Banned Items

Certain items considered to be problematic are prohibited or regulated on many mass transit systems. These include firearms and other weapons (unless licensed to carry), explosives, flammable items, or hazardous chemicals and substances.

Many systems prohibit live animals, but allow those that are in carrying cases or other closed containers. Additionally, service animals for the blind or disabled are permitted.

Some systems prohibit items of a large size that may take up a lot of space, such as non-folding bicycles. But more systems in recent years have been permitting passengers to bring bikes.

In Sydney, it is illegal to carry spray cans or permanent markers on public transport, as they can be used to vandalise the vehicles and stations. This rule also applies to sharp instruments that could damage the train, such as screwdrivers that could be used to make "scratchitti", a form of vandalism where tags are carved into a window.

Other Regulations

Many systems have regulations against behavior deemed to be unruly or otherwise disturbing to other passengers. In such cases, it is usually at the discretion of the operator, police officers, or other transit employees to determine what behaviors fit this description.

Some systems have regulations against photography or videography of the system's vehicles, stations, or other property. Those seen holding a mobile phone in a manner consistent with photography are considered to be suspicious for breaking this rule. This is another issue that is challenged in the courts in the United States as a "Free Speech" issue. Almost all riders are equipped with cell phones which can take pictures or record what is happening on the lines. Riders are able to record the actions of transit police and transportation system employees.

Sleeping

In the era when long distance trips took several days, sleeping accommodations were an essential part of transportation. (On land, the lodging involved was often part of the infrastructure: the inn or ryokan, which did not move, sheltered travelers. People also slept on ships at sea.) Today, most airlines, inter-city trains and coaches offer reclining seats and many provide pillows and blankets for overnight travelers. Better sleeping arrangements are commonly offered for a premium fare and include sleeping cars on overnight trains, larger private cabins on ships and airplane seats that convert into beds. Budget-conscious tourists sometimes plan their trips using overnight train or bus trips in lieu of paying for a hotel. The ability to get additional sleep on the way to work is attractive to many commuters using public transport.

Because night trains or coaches can be cheaper than motels, homeless persons often use these as overnight shelters, as with the famous Line 22 ("Hotel 22") in Silicon Valley. Specifically, a local transit route with a long overnight segment and which accepts inexpensive multi-use passes will acquire a reputation as a "moving hotel" for people with limited funds. Most transportation agencies actively discourage this. For this and other reasons passengers are often required to exit the vehicle at the end of the line; they can board again in the same or another vehicle, after some waiting. Even a low fare in some cases often deters the poorest individuals, including homeless people.

Ramp Meter

A ramp meter, ramp signal or metering light is a device, usually a basic traffic light or a two-section signal (red and green only, no yellow) light together with a signal controller that regulates the flow of traffic entering freeways according to current traffic conditions. It is the use of traffic signals at freeway on-ramps to manage the rate of automobiles entering the freeway. Ramp metering systems have proved to be successful in decreasing traffic congestion and improving driver safety.

Metered ramp on I-894 in the Milwaukee area.

A Portland, Oregon ramp meter

Ramp meters are claimed to reduce congestion (increase speed and volume) on freeways by reducing demand and by breaking up platoons of cars. Two variations of demand reduction are commonly cited; one being access rate, the other diversion. Some ramp meters are designed and programmed to operate only at times of peak travel demand; such meters are turned off during off-peak times allowing traffic to merge onto the freeway without stopping. Other ramp meters are designed to operate continuously, only being turned off for maintenance or repairs.

Types

Some metered ramps have bypass lanes for high-occupancy vehicles, allowing carpoolers and buses to skip the queue and get directly on the highway. In other places such as Northern California, carpool lanes are still metered, but the queue is typically shorter in comparison to regular lanes.

Meters often only operate in rush hour periods. Some ramp meters have only one lane of traffic at the signal; others may have two or more lanes of traffic. Generally, meters with multiple lanes only give one lane the green light at a time. In one common configuration, each entrance lane has two signals; a red-yellow-green signal perched overhead over each lane (or mounted high on a pole for a single lane), and a two-phase lamp mounted low on a pole next to the stop line.

The overhead lights are for cars approaching the metering point; the low-mounted two-phase lights are intended to be used by the vehicle at the front of the queue. In normal operation of the ramp meters, only the red and green lamps are used. However, when ramp metering is about to be enabled, the overhead lamps may show flashing or solid yellow to warn drivers to prepare to stop. (Once ramp metering is turned on, there is no further need for the yellow lamp.) In California, some meters allow two or three cars to proceed on a green light. These meters use red-yellow-green signals on both the upper and lower mounts on the pole, and operate in a standard green-yellow-red fashion.

In Ontario, the ramp meter lights are always green when there are no restrictions in place for traffic to proceed.

The sophistication and extent of a ramp metering system should be based on the amount of improvement desired, existing traffic conditions, installation costs, and the continuing resource requirements that are necessary to operate and maintain the system effectively. The simplest form of control is a fixed time operation. It performs the basic functions of breaking up platoons into single-vehicle entries and setting an upper limit on the flow rates that enter the freeway. Presence and passage detectors may be installed on the ramp to actuate and terminate the metering cycles, but the metering rate is based on average traffic conditions at a particular ramp at a particular time. This type of operation provides the benefits associated with accident reductions, but is not as effective in regulating freeway volumes because there is no input about mainline traffic. Pre-timed control can be implemented on any number of ramps, and is often implemented as an initial operating strategy until individual ramps can be incorporated into a traffic responsive system. The next level of control, traffic responsive, establishes metering rates based on actual freeway conditions. The local traffic responsive approach utilizes detectors and a micro-processor to determine the mainline flow in the immediate vicinity of the ramp and the ramp demand to select an appropriate metering rate. Traffic responsive control also permits ramp metering to be used to help manage demand when incidents occur on the freeway, i.e. reduce the metering rate at ramps upstream of the incident and increase the rate at ramps downstream. System-wide control is a form of traffic responsive control but operates on the basis of total freeway conditions. Centralized computer controlled systems can handle numerous ramps in a traffic responsive scheme and feature multiple control programs and overrides. Control strategies can also be distributed among individual ramps. A significant feature of system control is interconnection that permits the metering rate at any ramp to be influenced by conditions at other locations. Denver showed that this type of control has significant benefits when properly applied.'

System control need not be limited to the freeway and its ramps. The concept of integrated traffic control combines or coordinates freeway and arterial street control systems to operate on the basis of corridor wide traffic conditions. The potential advantages of integrated control include reduced installation and operating costs, corridor wide surveillance, better motorist information, and quicker and coordinated use of all of the control elements (meters, signals, signs, etc.) in response to real time traffic conditions. Simulation results from one study showed that, during an incident, coordination of arterial traffic signals and ramp meters can improve the traffic performance of a corridor.

Ramp Metering Signal Controls

Ramp meter signals are set according to the current traffic conditions on the road. Detectors (generally an induction loop) are installed in the road, both on the ramp and on the main road which measure and calculate the traffic flow, speed and occupancy levels. These are then used to alter the number of vehicles that can leave the ramp. The more congested the main carriageway the fewer vehicles are allowed to leave the ramp, this is effected by giving longer red times to the traffic signals.

Much research is currently being carried out into the most appropriate algorithms for controlling ramp meter signals. Some algorithms that are in use or have been evaluated are ALINEA, demand control and fuzzy algorithms.

Demand Control Algorithms

The demand control algorithms are examples of feed-forward control. One version of the demand control algorithm is the RWS strategy used in the Netherlands. In this algorithm the number of vehicles that the signals allow off the ramp is calculated as the difference between the flow before the ramp and the pre-specified capacity of the road.

Ramp metering in North America

This first application involved a police officer who would stop traffic on an entrance ramp and release vehicles one at a time at a predetermined rate, so that the objectives of safer and smoother merging onto the freeway traffic was easier without disrupting the mainline flows.

Ramp metering was first implemented in 1963 on the Eisenhower Expressway (Interstate 290) in Chicago by Adolf D. May, now a UC Berkeley professor. Since then ramp-meters have been systematically deployed in many urban areas including Los Angeles; San Diego; Sacramento; the San Francisco Bay Area; Fresno; Philadelphia, Pennsylvania; Seattle; Denver; Phoenix; Las Vegas; Salt Lake City; Portland; Minneapolis-St. Paul; Milwaukee; Columbus; Cincinnati; Houston; Atlanta; Miami; Washington, DC (only along Interstate 395 and Interstate 66 in Arlington County, Virginia); Kansas City, Missouri; and along the Queen Elizabeth Way in Mississauga, Ontario (Toronto-bound ramps from Cawthra Road, Hurontario Street, Mississauga Road, Erin Mills Parkway, Winston Churchill Boulevard, Ford Drive) Canada since the 1970s.

Ramp meters are commonplace in the New York City, Los Angeles, San Francisco, Chicago, Seattle, Phoenix, Houston, Atlanta, Milwaukee, Columbus, and Minneapolis-St. Paul metropolitan areas, and they are also found in more than two dozen smaller metropolitan areas. In the New York City metro area, locals refer to ramp meters as "merge lights" and in Houston they're known as "flow signals."

Ramp meters have been withdrawn after initial introduction in several cities, including Dallas, San Antonio, and Austin, Texas. Disused metering signals can still be found along some parkways surrounding New York City and Detroit, as well as on one ramp to Interstate 64/U.S. Route 40 in St. Louis, MO, on the WB entrance ramp from McCausland Ave., which will be removed as part of the New I-64 project in winter 2008. Although deactivated shortly after they were added, ramp meters have been reactivated at select interchanges of Interstate 476 in suburban Philadelphia.

A Milwaukee, Wisconsin ramp meter

Ramp meters were installed along Interstate 435 in Overland Park, Kansas and Kansas City, Missouri in 2009, and are scheduled to go into effect on November 24, 2009.

Ramp meters in Mississauga, Ontario are designed in such a way so that if the queue waiting to enter the QEW grows to the point where it may back up onto city streets, the meter is lifted and all traffic entering the highway is able to move freely without waiting for the meter. The meter goes back into service once the ramp queue is reduced to a reasonable level. While this method may increase congestion on the highway itself, it has the benefit of keeping city arterials free of stopped traffic waiting in queue. Ramp queues are usually quite short, lasting only 5–6 seconds on average before the driver may continue to the freeway.

Minneapolis–Saint Paul Ramp Meter Experiment

In 2000, a $650,000 experiment was mandated by the Minnesota State Legislature in response to citizen complaints and the efforts of State Senator Dick Day . The study involved shutting off all 433 ramp meters in the Minneapolis-St. Paul area for eight weeks to test their effectiveness. The study was conducted by Cambridge Systematics and concluded that when the ramp meters were turned off freeway capacity decreased by 9%, travel times increased by 22%, freeway speeds dropped by 7% and crashes increased by 26%. However, ramp meters remain controversial, and the Minnesota State Department of Transportation has developed new ramp control strategies. Fewer meters are activated during the course of a normal day than prior to the 2000 study, some meters have been removed, timing has been altered so that no driver waits more than four minutes in ramp queue, and vehicles are not allowed to back up onto city streets.

Mainline Metering

A mainline meter throttles traffic flow from one segment of a highway to the next by directly metering the highway's traffic. Such a scheme is typically implemented in specialized situations such as bridges and tunnels. A mainline meter was installed at the San Francisco–Oakland Bay Bridge toll plaza in the early 1970s. Similar mainline meters have also been installed downstream from the toll plazas at two other San Francisco Bay crossings, the San Mateo Bridge and the Dumbarton Bridge. However, these mainline meters have not yet been activated (as of September 2006). A mainline meter also exists on California State Route 125 southbound at its junction with Interstate 8 in La Mesa, California.

Ramp Metering in Europe

Ramp metering has been installed in several countries in Europe, including the United Kingdom, Germany and the Netherlands. A research project, *EURAMP - European Ramp Metering Project*, funded by the European Union was completed in March 2007. The EURAMP Project Deliverables included information about the results of ramp metering in a number of locations and situations, and whether they were helpful in those situations, and a *Handbook of ramp metering*.

United Kingdom

The first trial in the UK was on the M6 J10 near Walsall in 1986. No more sites were developed for the next two decades until a second 'pilot' study in 2006 by the Highways Agency (HA) concluded that ramp metering provides a net benefit under certain conditions - generally more congested junctions. A Summary Report by the HA, dated November 2007, includes an overview of the background and history, international experience, limitations, system operation, algorithms and implementation of ramp metering. In its conclusion it "envisaged ramp metering will be deployed more widely in the coming period." Ramp metering was then introduced widely in England - Phase 1 involved the implementation of approximately 30 sites and was completed by 2008. Phase 2 followed and as of March 2011 there are 88 Ramp Metering sites on the 4,500 miles of strategic highways operated and maintained by the HA.

Ramp Metering on the A19 in Middlesbrough.

The Netherlands

The first ramp metering in the Netherlands was introduced in 1989. Ramp metering is being introduced more widely in the Netherlands after a pilot study by the AVV Transport Research Centre which concluded that ramp metering can provide a small benefit for the traffic flow on the highway, leading to a higher capacity. Ramp meters can also contribute to decreasing 'rat running'. By 2006 50 ramp meters were installed. This number increases by 4 to 5 each year.

Germany

Ramp metering has been implemented on Autobahns in several areas in Germany, including the Rhine-Ruhr area, Munich, and Hamburg.

Italy

Ramp metering has been implemented on Tangenziale di Venezia (A57) as temporary solution for the increased traffic before the definitive solution (building of the Passante di Mestre).

Ramp Metering Elsewhere

Japan

Ramp metering is being installed in Japan in the next few years to keep the flow of traffic moving in Japan. There are plans to install ramp meters on every on-ramp in the Japan motorway system.

Australia

The largest Ramp metering network in the country, is used by the city with the largest Freeway and Road network - Melbourne (managed and controlled through VicRoads); including the Eastern Freeway and most of all, on the entire inner-city M1 route which includes the Monash Freeway, the CityLink Tollway, the West Gate Freeway and the metropolitan section (south of the beginning of the Monash Freeway) for the Princes Freeway. There are also various Ramp Meters on the inner-city section of the Calder Freeway. Ramp metering is used to regulate access to a number of major roads in Sydney, including: M4 Western Motorway (Wallgrove Road on-ramp); the M5 East motorway (Kingsgrove Road on-ramp); and the citywest Link to Anzac Bridge Brisbane's Pacific Motorway also uses Ramp metering on some on-ramps. On most motorways, ramp metering is activated when sensors indicate that traffic is heavy, however, some motorways without sensors use time-based activation.

The recent M1 Upgrade in Melbourne has installed 62 ramp meters that are coordinated using the HERO suite of algorithms developed by Markos Papageorgiou and Associates from the Technical University of Crete. The system has been built on the STREAMS platform and utilises the state-of-the-art ITS architecture. All the ramps can be linked when required to resolve motorway bottlenecks before they emerge. The results of a recent trial improved capacity by 9% over the previous fixed time ramp metering system, average speeds increased by 20kmh and traffic throughput at bottleneck locations can be reliably maintained around 2200 PCE per lane (note the M1 is major freight route to the Port of Melbourne and has a very high heavy articulated vehicle mix. The HERO system takes real time data every 20 seconds from the motorway, ramps and arterial road in order determine the best signal timing for the next 20 seconds. The data detection system comprises Sensys detectors in every freeway lane at 500m spacings with a minimum detectors at three locations on each ramp including the freeway entrance with the arterial road. The system also manages the arterial road interface with the freeway, balances ramp queues and delays across ramps, and is capable of managing bottlenecks 3–4 km downstream of a ramp entrance. The system is also supplemented by real time travel time information to key destinations and incident and congestion information displayed on specially designed full colour VMS on the approaches to the freeway entrance ramps. This information provides sufficient advice for motorists to determine whether or not to use the freeway during incidents etc. The system also provides dynamic ramp closure in the event of a major incident.

New Zealand

Ramp metering on North Western to Northern connection in Auckland.

Auckland has currently 84 ramp meters across the Southern, South Western, Northern and North Western motorways making it the largest Southern Hemisphere ramp metering system. Ramp metering was installed Auckland-wide after a successful trial on Mahunga Drive in 2004, before the Mangere Bridge.

Traffic data collected from 25 ramp metering sites in 2007 (before ramp metering deployment) and 2009 (after) shows an average 25% improvement in both congestion duration and traffic speed as well as an 8% increase in traffic throughput. The data also shows an average reduction in crashes of 22%. This performance and safety data translates into estimated benefits of US$1.6M per ramp metering site per year.

The system controlling the ramps promotes the traditional coordination among on-ramps as well as real-time integration with traffic signals on the adjacent arterial network allowing the whole road network to be managed as a single integrated network. For example, when motorway incidents adversely impact the adjacent arterial roads, an automatic response to the arterial traffic signals can be triggered to mitigate the impacts of the incident and vice versa. Recurrent and excessive traffic queues at on-ramp and off-ramp can also be managed in an integrated way in real-time. This integrated management is possible in Auckland because the same adaptive SCATS system controls both arterial traffic lights and motorway ramp meters.

The term *Ramp Signalling* rather than *Ramp Metering* is purposefully adopted in New Zealand as a user-oriented name.

South Africa

Ramp meters were, for a while, installed on the Samrand South bound, Old Johannesburg South bound and on New Road North and South bound interchanges on the N1 Ben Schoeman highway. The ramp metering was part of the Intelligent Transport System launched in October 2007 to aid traffic flow between Johannesburg and Pretoria.

Has also been installed on the north bound on ramp from Blue Lagoon to the M4 Highway in Durban since early 2007

Taiwan

Freeways in Taiwan use ramp meters during peak hours since 1993. Traffic enforcement cameras are deployed to deter running the red lights, but a bus lane at Taipei Interchange from northbound Chongqing North Road to southbound National Highway No. 1 in northern Datong District, Taipei allows buses and properly indicated emergency vehicles to bypass the traffic control imposed by the ramp meters.

References

- Needham, Joseph. (1986). Science and Civilization in China: Volume 4, Physics and Physical Technology, Part 3, Civil Engineering and Nautics. Taipei: Caves Books, Ltd. ISBN 0-521-07060-0
- Faulds, Ann; Craggs, Trudi & Saunders, John (31 January 2008). Chapter 4: The Definition of a Road?. Scottish Roads Law (2nd ed.). Practical Law Company. Retrieved 21 March 2014.
- "Rapid transit". Merriam-Webster. Retrieved 27 February 2008.; "Metro". International Association of Public Transport. Archived from the original on 2013-06-27. Retrieved 27 February 2008.
- "Transportation Invest In Our Future – Surface Transportation Policy Recommendations". Transportation1.org. Archived from the original on 2013-03-10. Retrieved 21 October 2011.
- "15-3 Length of Transport Routes at Year-end by Region". China National Bureau of Statististcs. 2007. Retrieved 12 January 2013.
- "Planning for Town Centres; Practice guidance on need, impact and the sequential approach" (PDF). Department for Communities and Local Government. December 2009. Retrieved March 26, 2012.
- "Transport Assessment; Guidelines for Development Proposals in Northern Ireland" (PDF). Department for Regional Development. November 9, 2006. Retrieved March 26, 2012.
- "Technical Guidance on Accessibility Planning in Local Transport Plans" (PDF). Local Transport Planning Network. Retrieved March 26, 2012.
- Barker, Kate (December 2006). "Barker Review of Land Use Planning" (PDF). Barker Review of Land Use Planning. Retrieved March 26, 2012
- "Canadian auto subsidies: Money for nothing? | vox – Research-based policy analysis and commentary from leading economists". Voxeu.org. 21 December 2008. Retrieved 21 October 2011.
- "Highway Construction Cost / Inflation Issues – Contract Admin – Construction – FHWA". Fhwa.dot.gov. 22 August 2011. Retrieved 21 October 2011.
- Victoria Transport Public Institute. "Victoria Transport Institute – Transportation Cost and Benefit Analysis". Vtpi.org. Retrieved 21 October 2011.
- "GAO-05-423SP Highlights of an Expert Panel: The Benefits and Costs of Highway and Transit Investments" (PDF). Retrieved 21 October 2011.

4

Intelligent Transportation Systems and Its Techniques

Artificial intelligence, computer simulation and real-time information on traffic tendencies all provide smarter alternatives to the current scenario of traffic regulation and management. Intelligent transportation systems are a complex of intelligent systems that can process and implement efficient transportation and regulation of vehicular traffic. Transportation and vehicular engineering is best understood in confluence with the major topics listed in the following chapter.

Intelligent Transportation System

Intelligent transportation systems (ITS) are advanced applications which, without embodying intelligence as such, aim to provide innovative services relating to different modes of transport and traffic management and enable various users to be better informed and make safer, more coordinated, and 'smarter' use of transport networks.

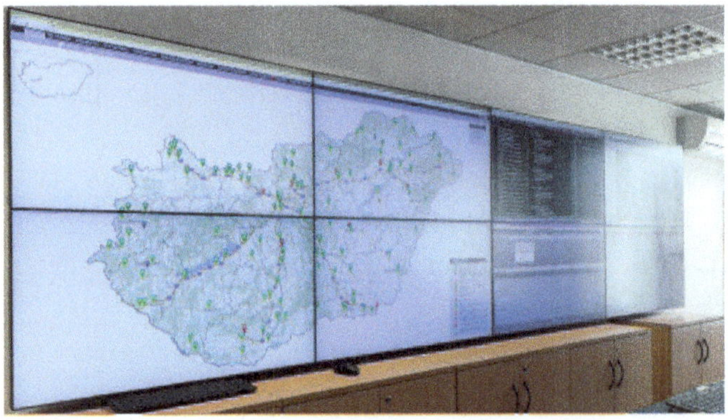

ITS graphical user interface displaying the Hungarian highway network and its data points

Although ITS may refer to all modes of transport, EU Directive 2010/40/EU (7 July 2010) defines ITS as systems in which information and communication technologies are applied in the field of road transport, including infrastructure, vehicles and users, and in traffic management and mobility management, as well as for interfaces with other modes of transport.

Background

Recent governmental activity in the area of ITS – specifically in the United States – is further motivated by an increasing focus on homeland security. Many of the proposed ITS systems also involve surveillance of the roadways, which is a priority of homeland security. Funding of many systems comes either directly through homeland security organisations or with their approval. Further,

ITS can play a role in the rapid mass evacuation of people in urban centers after large casualty events such as a result of a natural disaster or threat. Much of the infrastructure and planning involved with ITS parallels the need for homeland security systems.

In the developing world, the migration from rural to urbanized habitats has progressed differently. Many areas of the developing world have urbanised without significant motorisation and the formation of suburbs. A small portion of the population can afford automobiles, but the automobiles greatly increase congestion in these multimodal transportation systems. They also produce considerable air pollution, pose a significant safety risk, and exacerbate feelings of inequities in the society. High-population density could be supported by a multimodal system of walking, bicycle transportation, motorcycles, buses, and trains.

Other parts of the developing world, such as China, India & Brazil remain largely rural but are rapidly urbanising and industrialising. In these areas a motorised infrastructure is being developed alongside motorisation of the population. Great disparity of wealth means that only a fraction of the population can motorise, and therefore the highly dense multimodal transportation system for the poor is cross-cut by the highly motorised transportation system for the rich.

Intelligent Transportation Technologies

Intelligent transport systems vary in technologies applied, from basic management systems such as car navigation; traffic signal control systems; container management systems; variable message signs; automatic number plate recognition or speed cameras to monitor applications, such as security CCTV systems; and to more advanced applications that integrate live data and feedback from a number of other sources, such as parking guidance and information systems; weather information; bridge de-icing (US deicing) systems; and the like. Additionally, predictive techniques are being developed to allow advanced modelling and comparison with historical baseline data. Some of these technologies are described in the following sections.

Wireless Communications

Traffic monitoring gantry with wireless communication dish antenna

Various forms of wireless communications technologies have been proposed for intelligent transportation systems.

Radio modem communication on UHF and VHF frequencies are widely used for short and long range communication within ITS.

Short-range communications of 350 m can be accomplished using IEEE 802.11 protocols, specifically WAVE or the Dedicated Short Range Communications standard being promoted by the Intelligent Transportation Society of America and the United States Department of Transportation. Theoretically, the range of these protocols can be extended using Mobile ad hoc networks or Mesh networking.

Longer range communications have been proposed using infrastructure networks such as WiMAX (IEEE 802.16), Global System for Mobile Communications (GSM), or 3G. Long-range communications using these methods are well established, but, unlike the short-range protocols, these methods require extensive and very expensive infrastructure deployment. There is lack of consensus as to what business model should support this infrastructure.

Auto Insurance companies have utilised ad hoc solutions to support eCall and behavioural tracking functionalities in the form of Telematics 2.0.

Computational Technologies

Recent advances in vehicle electronics have led to a move towards fewer, more capable computer processors on a vehicle. A typical vehicle in the early 2000s would have between 20 and 100 individual networked microcontroller/Programmable logic controller modules with non-real-time operating systems. The current trend is toward fewer, more costly microprocessor modules with hardware memory management and real-time operating systems. The new embedded system platforms allow for more sophisticated software applications to be implemented, including model-based process control, artificial intelligence, and ubiquitous computing. Perhaps the most important of these for Intelligent Transportation Systems is artificial intelligence.

Floating Car Data/floating Cellular Data

RFID E-ZPass reader attached to the pole and its antenna (right) used in traffic monitoring in New York City by using vehicle re-identification method

"Floating car" or "probe" data collection is a set of relatively low-cost methods for obtaining travel time and speed data for vehicles travelling along streets, highways, motorways (freeways), and other transport routes. Broadly speaking, four methods have been used to obtain the raw data:

- Triangulation method. In developed countries a high proportion of cars contain one or more mobile phones. The phones periodically transmit their presence information to the mobile phone network, even when no voice connection is established. In the mid-2000s, attempts were made to use mobile phones as anonymous traffic probes. As a car moves, so does the signal of any mobile phones that are inside the vehicle. By measuring and analysing network data using triangulation, pattern matching or cell-sector statistics (in an anonymous format), the data was converted into traffic flow information. With more congestion, there are more cars, more phones, and thus, more probes. In metropolitan areas, the distance between antennas is shorter and in theory accuracy increases. An advantage of this method is that no infrastructure needs to be built along the road; only the mobile phone network is leveraged. But in practice the triangulation method can be complicated, especially in areas where the same mobile phone towers serve two or more parallel routes (such as a motorway (freeway) with a frontage road, a motorway (freeway) and a commuter rail line, two or more parallel streets, or a street that is also a bus line). By the early 2010s, the popularity of the triangulation method was declining.

- Vehicle re-identification. Vehicle re-identification methods require sets of detectors mounted along the road. In this technique, a unique serial number for a device in the vehicle is detected at one location and then detected again (re-identified) further down the road. Travel times and speed are calculated by comparing the time at which a specific device is detected by pairs of sensors. This can be done using the MAC addresses from Bluetooth or other devices, or using the RFID serial numbers from Electronic Toll Collection (ETC) transponders (also called "toll tags").

- GPS based methods. An increasing number of vehicles are equipped with in-vehicle satnav/GPS (satellite navigation) systems that have two-way communication with a traffic data provider. Position readings from these vehicles are used to compute vehicle speeds. Modern methods may not use dedicated hardware but instead Smartphone based solutions using so called Telematics 2.0 approaches.

- Smartphone-based rich monitoring. Smartphones having various sensors can be used to track traffic speed and density. The accelerometer data from smartphones used by car drivers is monitored to find out traffic speed and road quality. Audio data and GPS tagging of smartphones enables identification of traffic density and possible traffic jams. This was implemented in Bangalore, India as a part of a research experimental system Nericell.

Floating car data technology provides advantages over other methods of traffic measurement:

- Less expensive than sensors or cameras
- More coverage (potentially including all locations and streets)
- Faster to set up and less maintenance
- Works in all weather conditions, including heavy rain

Sensing Technologies

Technological advances in telecommunications and information technology, coupled with ultramodern/state-of-the-art microchip, RFID (Radio Frequency Identification), and inexpensive intelligent beacon sensing technologies, have enhanced the technical capabilities that will facilitate motorist safety benefits for intelligent transportation systems globally. Sensing systems for ITS are vehicle- and infrastructure-based networked systems, i.e., Intelligent vehicle technologies. Infrastructure sensors are indestructible (such as in-road reflectors) devices that are installed or embedded in the road or surrounding the road (e.g., on buildings, posts, and signs), as required, and may be manually disseminated during preventive road construction maintenance or by sensor injection machinery for rapid deployment. Vehicle-sensing systems include deployment of infrastructure-to-vehicle and vehicle-to-infrastructure electronic beacons for identification communications and may also employ video automatic number plate recognition or vehicle magnetic signature detection technologies at desired intervals to increase sustained monitoring of vehicles operating in critical zones.

Inductive Loop Detection

Saw cut loop detectors for vehicle detection buried in the pavement at this intersection as seen by the rectangular shapes of loop detector sealant at the bottom part of this picture.

Inductive loops can be placed in a roadbed to detect vehicles as they pass through the loop's magnetic field. The simplest detectors simply count the number of vehicles during a unit of time (typically 60 seconds in the United States) that pass over the loop, while more sophisticated sensors estimate the speed, length, and class of vehicles and the distance between them. Loops can be placed in a single lane or across multiple lanes, and they work with very slow or stopped vehicles as well as vehicles moving at high speed.

Video Vehicle Detection

Traffic-flow measurement and automatic incident detection using video cameras is another form of vehicle detection. Since video detection systems such as those used in automatic number plate recognition do not involve installing any components directly into the road surface or roadbed, this type of system is known as a "non-intrusive" method of traffic detection. Video from cameras is fed into processors that analyse the changing characteristics of the video image as vehicles pass. The cameras are typically mounted on poles or structures above or adjacent to the roadway. Most video detection systems require some initial configuration to "teach" the processor the baseline background image. This usually involves inputting known measurements such as the distance between lane lines or the height of the camera above the roadway. A single video detection processor can detect traffic simultaneously from one to eight cameras, depending on the brand and model. The typical output from a video detection system is lane-by-lane vehicle speeds, counts, and lane occupancy readings. Some systems provide additional outputs including gap, headway, stopped-vehicle detection, and wrong-way vehicle alarms.

Bluetooth Detection

Bluetooth is an accurate and inexpensive way to measure travel time and make origin and destination analysis. Bluetooth is a wireless standard used to communicate between electrondresses from Bluetooth devices in passing vehicles. If these sensors are interconnected they are able to calculate travel time and provide data for origin and destination matrices. Compared to other traffic measurement technologies, Bluetooth measurement has some differences:

- Accurate measurement points with absolute confirmation to provide to the second travel times.
- Is non-intrusive, which can lead to lower-cost installations for both permanent and temporary sites.
- Is limited to how many Bluetooth devices are broadcasting in a vehicle so counting and other applications are limited.
- Systems are generally quick to set up with little to no calibration needed.

Since Bluetooth devices become more prevalent on board vehicles and with more portable electronics broadcasting, the amount of data collected over time becomes more accurate and valuable for travel time and estimation purposes.

Audio Detection

It is also possible to measure traffic density on a road using the Audio signal that consists of the cumulative sound from tire noise, engine noise, engine-idling noise, honks and air turbulence noise. A roadside-installed microphone picks up the audio that comprises the various vehicle noise and Audio signal processing techniques can be used to estimate the traffic state. The accuracy of such a system compares well with the other methods described above.

Information Fusion from Multiple Traffic Sensing Modalities

The data from the different sensing technologies can be combined in intelligent ways to determine the traffic state accurately. A Data fusion based approach that utilizes the road side collected acoustic, image and sensor data has been shown to combine the advantages of the different individual methods.

Intelligent Transportation Applications

Emergency Vehicle Notification Systems

The in-vehicle eCall is generated either manually by the vehicle occupants or automatically via activation of in-vehicle sensors after an accident. When activated, the in-vehicle eCall device will establish an emergency call carrying both voice and data directly to the nearest emergency point (normally the nearest E1-1-2 Public-safety answering point, PSAP). The voice call enables the vehicle occupant to communicate with the trained eCall operator. At the same time, a minimum set of data will be sent to the eCall operator receiving the voice call.

The minimum set of data contains information about the incident, including time, precise location, the direction the vehicle was traveling, and vehicle identification. The pan-European eCall aims to be operative for all new type-approved vehicles as a standard option. Depending on the manufacturer of the eCall system, it could be mobile phone based (Bluetooth connection to an in-vehicle interface), an integrated eCall device, or a functionality of a broader system like navigation, Telematics device, or tolling device. eCall is expected to be offered, at earliest, by the end of 2010, pending standardization by the European Telecommunications Standards Institute and commitment from large EU member states such as France and the United Kingdom.

Congestion pricing gantry at North Bridge Road, Singapore.

The EC funded project SafeTRIP is developing an open ITS system that will improve road safety and provide a resilient communication through the use of S-band satellite communication. Such platform will allow for greater coverage of the Emergency Call Service within the EU.

Automatic Road Enforcement

Automatic speed enforcement gantry or "*Lombada Eletrônica*" with ground sensors at Brasilia, D.F.

A traffic enforcement camera system, consisting of a camera and a vehicle-monitoring device, is used to detect and identify vehicles disobeying a speed limit or some other road legal requirement and automatically ticket offenders based on the license plate number. Traffic tickets are sent by mail. Applications include:

- Speed cameras that identify vehicles traveling over the legal speed limit. Many such devices use radar to detect a vehicle's speed or electromagnetic loops buried in each lane of the road.
- Red light cameras that detect vehicles that cross a stop line or designated stopping place while a red traffic light is showing.
- Bus lane cameras that identify vehicles traveling in lanes reserved for buses. In some jurisdictions, bus lanes can also be used by taxis or vehicles engaged in car pooling.
- Level crossing cameras that identify vehicles crossing railways at grade illegally.
- Double white line cameras that identify vehicles crossing these lines.
- High-occupancy vehicle lane cameras that identify vehicles violating HOV requirements.
- Turn cameras at intersections where specific turns are prohibited on red. This type of camera is mostly used in cities or heavy populated areas.

Variable Speed Limits

Example variable speed limit sign in the United States.

Recently some jurisdictions have begun experimenting with variable speed limits that change with road congestion and other factors. Typically such speed limits only change to decline during poor conditions, rather than being improved in good ones. One example is on Britain's M25 motorway, which circumnavigates London. On the most heavily traveled 14-mile (23 km) section (junction 10 to 16) of the M25 variable speed limits combined with automated enforcement have been in force since 1995. Initial results indicated savings in journey times, smoother-flowing traffic, and a fall in the number of accidents, so the implementation was made permanent in 1997. Further trials on the M25 have been thus far proven inconclusive.

Collision Avoidance Systems

Japan has installed sensors on its highways to notify motorists that a car is stalled ahead.

Dynamic Traffic Light Sequence

A 2008 paper was written about using RFID for dynamic traffic light sequences. It circumvents or avoids problems that usually arise with systems that use image processing and beam interruption techniques. RFID technology with appropriate algorithm and database were applied to a multi-vehicle, multi-lane and multi-road junction area to provide an efficient time management scheme. A dynamic time schedule was worked out for the passage of each column. The simulation showed the dynamic sequence algorithm could adjust itself even with the presence of some extreme cases. The paper said the system could emulate the judgment of a traffic police officer on duty, by considering the number of vehicles in each column and the routing proprieties.

Cooperative Systems on the Road

Communication cooperation on the road includes car-to-car, car-to-infrastructure, and vice versa. Data available from vehicles are acquired and transmitted to a server for central fusion and processing. These data can be used to detect events such as rain (wiper activity) and congestion (frequent braking activities). The server processes a driving recommendation dedicated to a single or a specific group of drivers and transmits it wirelessly to vehicles. The goal of cooperative systems is to use and plan communication and sensor infrastructure to increase road safety. The definition of cooperative systems in road traffic is according to the European Commission:

> "Road operators, infrastructure, vehicles, their drivers and other road users will cooperate to deliver the most efficient, safe, secure and comfortable journey. The vehicle-vehicle and vehicle-infrastructure co-operative systems will contribute to these objectives beyond the improvements achievable with stand-alone systems."

ITS World Congress is a world-wide annual event to promote and showcase ITS technologies. ERTICO – ITS Europe, ITS America and ITS AsiaPacific work closely together in the preparation of the annual ITS World Congress and Exhibition attracting over 8,000 people. Each year the event takes place in a different region (Europe, Americas or Asia-Pacific).

The first ITS World Congress was held in Paris in 1994 and it has become the most important annual ITS event. Subsequent events have been held in:

- 1995 Yokohama
- 1996 Orlando
- 1997 Berlin

- 1998 Seoul
- 1999 Toronto
- 2000 Turin
- 2001 Sydney
- 2002 Chicago
- 2003 Madrid
- 2004 Nagoya
- 2005 San Francisco
- 2006 London
- 2007 Beijing
- 2008 New York
- 2009 Stockholm
- 2010 Busan
- 2011 Orlando
- 2012 Vienna
- 2013 Tokyo
- 2014 Detroit
- 2015 Bordeaux

Future Congress are scheduled for Melbourne 2016, Montreal 2017, Copenhagen 2018, Singapore 2019.

Europe

The Network of National ITS Associations is a grouping of national ITS interests formed in order to ensure that ITS knowledge and information is transmitted to all actors at the local and national level – such as small and medium-sized companies – and support ITS promotion from the ground up. It was officially launched 7 October 2004 in London. The Network currently consists of 27 member organisations. The Network Secretariat is at ERTICO – ITS Europe.

ERTICO – ITS Europe is a multi-sector, public/private partnership pursuing the development and deployment of Intelligent Transport Systems and Services (ITS). They connect public authorities, industry players, infrastructure operators, users, national ITS associations and other organisations together and work to bring "Intelligence into Mobility". The ERTICO work programme focuses on initiatives to improve transport safety, security and network efficiency whilst taking into account measures to reduce environmental impact. Our vision is of a future transport system working towards zero accidents, zero delays and fully informed people, where services are affordable and seamless, the environment is protected, privacy is respected and security is provided.

United States

In the United States, each state has an Intelligent Transportation Systems (ITS) chapter that holds a yearly conference to promote and showcase ITS technologies and ideas. Representatives from each Department of Transportation (state, cities, towns, and counties) within the state attend this conference.

Automotive Navigation System

An automotive navigation system is part of the automobile controls or a third party add-on used to find direction in an automobile. It typically uses a satellite navigation device to get its position data which is then correlated to a position on a road. When directions are needed routing can be calculated. On the fly traffic information can be used to adjust the route.

Dead reckoning using distance data from sensors attached to the drivetrain, a gyroscope and an accelerometer can be used for greater reliability, as GPS signal loss and/or multipath can occur due to urban canyons or tunnels.

History

Automotive navigation systems represent a convergence of a number of diverse technologies many of which have been available for many years, but were too costly or inaccessible. Limitations such as batteries, display, and processing power had to be overcome before the product became commercially viable. Etak made an early system that used map-matching to improve on dead reckoning instrumentation. Digital map information was stored on standard cassette tapes.

- 1966: General Motors Research (GMR) was working on a non-satellite-based navigation and assistance system called DAIR (Driver Aid, Information & Routing). After initial tests GM found that it wasn't a scalable or practical way to provide navigation assistance. Decades later, however, the concept would be reborn as OnStar.
- 1980: Electronic Auto Compass with new mechanism on the Toyota Crown.
- 1981: navigation computer on the Toyota Celica. (NAVICOM)
- 1987: Toyota introduced the World's first CD-ROM-based navigation system on the Toyota Crown.
- 1990: Mazda Eunos Cosmo became the first car with built-in GPS-navigation system
- 1991: Toyota introduced GPS car navigation on the Toyota Soarer.
- 1992: Voice assisted GPS navigation system on the Toyota Celsior.
- 1995: Oldsmobile introduced the first GPS navigation system available in a United States production car, called GuideStar.
- 1995: Device called "Mobile Assistant" or short, MASS, produced by Munich-based compa-

ny ComRoad AG, won the title "Best Product in Mobile Computing" on CeBit by magazine Byte. It offered turn-by-turn navigation via wireless internet connection, with both GPS and speed sensor in the car.

- 1995: Renault Safrane introduce first European CD based navigation system (CARMINAT Philips - Renault), big color display modern interface and 2D map view.
- 1995: BMW 7 series E38 second European model with the Sat Nav, using pretty same software and hardware (Philips - Renault) like Renault Safrane, build in magnetometer, more features available from the menu like: AUX ventilation, TV or Phone
- 1997: Navigation system using Differential GPS developed as a factory-installed option on the Toyota Prius
- 1998: First DVD-based navigation system introduced on the Toyota Progres
- 2000: The United States made a more accurate GPS signal available for civilian use.
- 2003: Toyota introduced the first Hard disk drive-based navigation system and the industry's first DVD-based navigation system with a built-in Electronic throttle control
- 2007: Toyota introduced Map on Demand, a technology for distributing map updates to car navigation systems, developed as the first of its kind in the world
- 2008: World's first navigation system-linked brake assist function and Navigation system linked to Adaptive Variable Suspension System (NAVI/AI-AVS) on Toyota Crown

Technology

The road database is a vector map. Street names or numbers and house numbers are encoded as geographic coordinates so that the user can find some desired destination by street address.

Points of interest (waypoints) are stored with their geographic coordinates. Formats are almost uniformly proprietary; there is no industry standard for satellite navigation maps, although some companies are trying to address this with SDAL and NDS.

Map data vendors such as Tele Atlas and Navteq create the base map in a Geographic Data Files format, but each electronics manufacturer compiles it in an optimized, usually proprietary format. GDF is not a CD standard for car navigation systems. GDF is used and converted onto the CD-ROM in the internal format of the navigation system.

CARiN Database Format (CDF) is a proprietary navigation map format created by Philips Car Systems (this branch was sold to Mannesman VDO, which became VDO/Dayton in 1998, then Siemens VDO in 2002, then Continental in 2007) and is used in a number of navigation-equipped vehicles. The 'CARiN' portmanteau is derived from Car Information and Navigation.

SDAL is a proprietary map format published by Navteq, who released it royalty free in the hope that it would become an industry standard for digital navigation maps. Vendors who used this format include:

- Microsoft
- Magellan
- Pioneer
- Panasonic
- Clarion
- InfoGation

The format has not been very widely adopted by the industry.

Navigation Data Standard (NDS)

The Navigation Data Standard (NDS) initiative, is an industry grouping of car manufacturers, navigation system suppliers and map data suppliers whose objective is the standardization of the data format used in car navigation systems, as well as allow a map update capability. The NDS effort began in 2004 and became a registered association in 2009. Standardization would improve interoperability, specifically by allowing the same navigation maps to be used in navigation systems from 20 manufacturers. Companies involved include BMW, Volkswagen, Daimler, Renault, ADIT, Aisin AW, Alpine Electronics, Navigon, Navis-AMS, Bosch, DENSO, Mitsubishi, Harman International Industries, Panasonic, PTV, Continental AG, Clarion, Navteq, Navinfo, TomTom and Zenrin.

Media

The road database may be stored in solid state read-only memory (ROM), optical media (CD or DVD), solid state flash memory, magnetic media (hard disk), or a combination. A common scheme is to have a base map permanently stored in ROM that can be augmented with detailed information for a region the user is interested in. A ROM is always programmed at the factory; the other media may be preprogrammed, downloaded from a CD or DVD via a computer or wireless connection (bluetooth, Wi-Fi), or directly used utilizing a card reader.

Some navigation device makers provide free map updates for their customers. These updates are often obtained from the vendor's website, which is accessed by connecting the navigation device to a PC.

Real-time Data

Some systems can receive and display information on traffic congestion using either TMC, RDS, or by GPRS/3G data transmission via mobile phones.

Integration and other functions

- The color LCD screens on some automotive navigation systems can also be used to display television broadcasts or DVD movies.

- A few systems integrate (or communicate) with mobile phones for hands-free talking and SMS messaging (i.e., using Bluetooth or Wi-Fi).
- Automotive navigation systems can include personal information management for meetings, which can be combined with a traffic and public transport information system.

Original Factory Equipment

Many vehicle manufacturers offer a GPS navigation device as an option in their vehicles. Customers whose vehicles did not ship with GPS can therefore purchase and retrofit the original factory-supplied GPS unit. In some cases this can be a straightforward "plug-and-play" installation if the required wiring harness is already present in the vehicle. However, with some manufacturers, new wiring is required, making the installation more complex.

The primary benefit of this approach an integrated and factory-standard installation. Many original systems also contain a gyrocompass or accelerometer and may accept input from the vehicle's speed sensors, thereby allowing them to navigate via dead reckoning when a GPS signal is temporarily unavailable. However, the costs can be considerably higher than other options.

SMS

Establishing points of interest in real-time and transmitting them via GSM cellular telephone networks using the Short Message Service (SMS) is referred to as Gps2sms. Some vehicles and vessels are equipped with hardware that is able to automatically send an SMS text message when a particular event happens, such as theft, anchor drift or breakdown. The receiving party (e.g., a tow truck) can store the waypoint in a computer system, draw a map indicating the location, or see it in an automotive navigation system.

Automatic Number Plate Recognition

Automatic number plate recognition (ANPR) is a technology that uses optical character recognition on images to read vehicle registration plates. It can use existing closed-circuit television, road-rule enforcement cameras, or cameras specifically designed for the task. ANPR is used by police forces around the world for law enforcement purposes, including to check if a vehicle is registered or licensed. It is also used for electronic toll collection on pay-per-use roads and as a method of cataloging the movements of traffic for example by highways agencies.

Automatic number plate recognition can be used to store the images captured by the cameras as well as the text from the license plate, with some configurable to store a photograph of the driver. Systems commonly use infrared lighting to allow the camera to take the picture at any time of the day. ANPR technology must take into account plate variations from place to place.

Concerns about these systems have centered on privacy fears of government tracking citizens' movements, misidentification, high error rates, and increased government spending. Critics have described it as a form of mass surveillance.

Other Names

ANPR is sometimes known by various other terms:

- Automatic license-plate recognition (ALPR)
- Automatic license-plate reader (ALPR)
- Automatic vehicle identification (AVI)
- Car plate recognition (CPR)
- License-plate recognition (LPR)
- Lecture automatique de plaques d'immatriculation (LAPI)
- Mobile license-plate reader (MLPR)
- Vehicle license-plate recognition (VLPR)

Development

ANPR was invented in 1976 at the Police Scientific Development Branch in the UK. Prototype systems were working by 1979, and contracts were awarded to produce industrial systems, first at EMI Electronics, and then at Computer Recognition Systems (CRS) in Wokingham, UK. Early trial systems were deployed on the A1 road and at the Dartford Tunnel. The first arrest through detection of a stolen car was made in 1981. However, ANPR did not become widely used until new developments in cheaper and easier to use software were pioneered during the 1990s. The collection of ANPR data for future use (*i.e.*, in solving then-unidentified crimes) was documented in the early 2000s. The first documented case of ANPR being used to help solve a murder occurred in November 2005, in Bradford, UK, where ANPR played a vital role in locating and subsequently convicting killers of Sharon Beshenivsky.

Components

The software aspect of the system runs on standard home computer hardware and can be linked to other applications or databases. It first uses a series of image manipulation techniques to detect, normalize and enhance the image of the number plate, and then optical character recognition (OCR) to extract the alphanumerics of the license plate. ANPR systems are generally deployed in one of two basic approaches: one allows for the entire process to be performed at the lane location in real-time, and the other transmits all the images from many lanes to a remote computer location and performs the OCR process there at some later point in time. When done at the lane site, the information captured of the plate alphanumeric, date-time, lane identification, and any other information required is completed in approximately 250 milliseconds. This information can easily be transmitted to a remote computer for further processing if necessary, or stored at the lane for later retrieval. In the other arrangement, there are typically large numbers of PCs used in a server farm to handle high workloads, such as those found in the London congestion charge project. Often in such systems, there is a requirement to forward images to the remote server, and this can require larger bandwidth transmission media.

Technology

The font on Dutch plates was changed to improve plate recognition.

ANPR uses optical character recognition (OCR) on images taken by cameras. When Dutch vehicle registration plates switched to a different style in 2002, one of the changes made was to the font, introducing small gaps in some letters (such as *P* and *R*) to make them more distinct and therefore more legible to such systems. Some license plate arrangements use variations in font sizes and positioning—ANPR systems must be able to cope with such differences in order to be truly effective. More complicated systems can cope with international variants, though many programs are individually tailored to each country.

The cameras used can existing road-rule enforcement or closed-circuit television cameras, as well as mobile units, which are usually attached to vehicles. Some systems use infrared cameras to take a clearer image of the plates.

In Mobile Systems

The Dubai police use ANPR cameras to monitor vehicles in front and either side of the patrol car

A Merseyside Police car equipped with mobile ANPR.

During the 1990s, significant advances in technology took automatic number plate recognition (ANPR) systems from limited expensive, hard to set up, fixed based applications to simple "point and shoot" mobile ones. This was made possible by the creation of software that ran on cheaper PC based, non-specialist hardware that also no longer needed to be given the pre-defined angles, direction, size and speed in which the plates would be passing the cameras field of view. Further scaled-down components at more cost-effective price points led to a record number of deployments by law enforcement agencies around the world. Smaller cameras with the ability to read license plates at higher speeds, along with smaller, more durable processors that fit in the trunks of police vehicles, allowed law enforcement officers to patrol daily with the benefit of license plate reading in real time, when they can interdict immediately.

Despite their effectiveness, there are noteworthy challenges related with mobile ANPRs. One of the biggest is that the processor and the cameras must work fast enough to accommodate relative speeds of more than 100 mph (160 km/h), a likely scenario in the case of oncoming traffic. This equipment must also be very efficient since the power source is the vehicle battery, and equipment must be small to minimize the space it requires.

Relative speed is only one issue that affects the camera's ability to actually read a license plate. Algorithms must be able to compensate for all the variables that can affect the ANPR's ability to produce an accurate read, such as time of day, weather and angles between the cameras and the license plates. A system's illumination wavelengths can also have a direct impact on the resolution and accuracy of a read in these conditions.

Installing ANPR cameras on law enforcement vehicles requires careful consideration of the juxtaposition of the cameras to the license plates they are to read. Using the right number of cameras and positioning them accurately for optimal results can prove challenging, given the various missions and environments at hand. Highway patrol requires forward-looking cameras that span multiple lanes and are able to read license plates at very high speeds. City patrol needs shorter range, lower focal length cameras for capturing plates on parked cars. Parking lots with perpendicularly parked cars often require a specialized camera with a very short focal length. Most technically advanced systems are flexible and can be configured with a number of cameras ranging from one to four which can easily be repositioned as needed. States with rear-only license plates have an additional challenge since a forward-looking camera is ineffective with oncoming traffic. In this case one camera may be turned backwards.

Algorithms

Steps 2, 3 and 4: The license plate is normalized for brightness and contrast, and then the characters are segmented to be ready for OCR.

There are seven primary algorithms that the software requires for identifying a license plate:

1. Plate localization – responsible for finding and isolating the plate on the picture.
2. Plate orientation and sizing – compensates for the skew of the plate and adjusts the dimensions to the required size.
3. Normalization – adjusts the brightness and contrast of the image.
4. Character segmentation – finds the individual characters on the plates.
5. Optical character recognition.
6. Syntactical/Geometrical analysis – check characters and positions against country-specific rules.
7. The averaging of the recognised value over multiple fields/images to produce a more reliable or confident result. Especially since any single image may contain a reflected light flare, be partially obscured or other temporary effect.

The complexity of each of these subsections of the program determines the accuracy of the system. During the third phase (normalization), some systems use edge detection techniques to increase the picture difference between the letters and the plate backing. A median filter may also be used to reduce the visual noise on the image.

Difficulties

Early ANPR systems were unable to read white or silver lettering on black background, as permitted on UK vehicles built prior to 1973.

Swedish license plate

Must be able to recognize international license plates as such.

There are a number of possible difficulties that the software must be able to cope with. These include:

- Poor file resolution, usually because the plate is too far away but sometimes resulting from the use of a low-quality camera.

- Blurry images, particularly motion blur.
- Poor lighting and low contrast due to overexposure, reflection or shadows.
- An object obscuring (part of) the plate, quite often a tow bar, or dirt on the plate.
- Read license plates that are different at the front and the back because of towed trailers, campers, etc.
- Vehicle lane change in the camera's angle of view during license plate reading.
- A different font, popular for vanity plates (some countries do not allow such plates, eliminating the problem).
- Circumvention techniques.
- Lack of coordination between countries or states. Two cars from different countries or states can have the same number but different design of the plate.

While some of these problems can be corrected within the software, it is primarily left to the hardware side of the system to work out solutions to these difficulties. Increasing the height of the camera may avoid problems with objects (such as other vehicles) obscuring the plate but introduces and increases other problems, such as the adjusting for the increased skew of the plate.

On some cars, tow bars may obscure one or two characters of the license plate. Bikes on bike racks can also obscure the number plate, though in some countries and jurisdictions, such as Victoria, Australia, "bike plates" are supposed to be fitted. Some small-scale systems allow for some errors in the license plate. When used for giving specific vehicles access to a barricaded area, the decision may be made to have an acceptable error rate of one character. This is because the likelihood of an unauthorized car having such a similar license plate is seen as quite small. However, this level of inaccuracy would not be acceptable in most applications of an ANPR system.

Imaging Hardware

At the front end of any ANPR system is the imaging hardware which captures the image of the license plates. The initial image capture forms a critically important part of the ANPR system which, in accordance to the garbage in, garbage out principle of computing, will often determine the overall performance.

Portable traffic enforcement system used by the Hungarian police.
The rows of infrared LEDs are visible on the right.

License plate capture is typically performed by specialized cameras designed specifically for the task, although new software techniques are being implemented that support any I.P.-based surveillance camera and increase the utility of ANPR for perimeter security applications. Factors which pose difficulty for license plate imaging cameras include the speed of the vehicles being recorded, varying level of ambient light, headlight glare and harsh environmental conditions. Most dedicated license plate capture cameras will incorporate infrared illumination in order to solve the problems of lighting and plate reflectivity.

Many countries now use license plates that are retroreflective. This returns the light back to the source and thus improves the contrast of the image. In some countries, the characters on the plate are not reflective, giving a high level of contrast with the reflective background in any lighting conditions. A camera that makes use of active infrared imaging (with a normal colour filter over the lens and an infrared illuminator next to it) benefits greatly from this as the infrared waves are reflected back from the plate. This is only possible on dedicated ANPR cameras, however, and so cameras used for other purposes must rely more heavily on the software capabilities. Further, when a full-colour image is required as well as use of the ANPR-retrieved details, it is necessary to have one infrared-enabled camera and one normal (colour) camera working together.

To avoid blurring it is ideal to have the shutter speed of a dedicated camera set to 1/1000 of a second. It is also important that the camera uses a global shutter, as opposed to rolling shutter, to assure that the taken images are distortion-free. Because the car is moving, slower shutter speeds could result in an image which is too blurred to read using the OCR software, especially if the camera is much higher up than the vehicle. In slow-moving traffic, or when the camera is at a lower level and the vehicle is at an angle approaching the camera, the shutter speed does not need to be so fast. Shutter speeds of 1/500 of a second can cope with traffic moving up to 40 mph (64 km/h) and 1/250 of a second up to 5 mph (8 km/h). License plate capture cameras can produce usable images from vehicles traveling at 120 mph (190 km/h).

To maximize the chances of effective license plate capture, installers should carefully consider the positioning of the camera relative to the target capture area. Exceeding threshold angles of incidence between camera lens and license plate will greatly reduce the probability of obtaining usable images due to distortion. Manufacturers have developed tools to help eliminate errors from the physical installation of license plate capture cameras.

Usage

Law Enforcement

Mobile ANPR cameras fitted to a New South Wales Police Force Highway Patrol vehicle.

Closed-circuit television cameras such as these can be used to take the images scanned by automatic number plate recognition systems

Australia

Several State Police Forces, and the Department of Justice (Victoria) use both fixed and mobile ANPR systems. The New South Wales Police Force Highway Patrol were the first to trial and use a fixed ANPR camera system in Australia in 2005. In 2009 they began a roll-out of a mobile ANPR system (known officially as MANPR) with three infrared cameras fitted to its Highway Patrol fleet. The system identifies unregistered and stolen vehicles as well as disqualified or suspended drivers as well as other 'persons of interest' such as persons having outstanding warrants.

Belgium

The city of Mechelen uses an ANPR system since September 2011 to scan all cars crossing the city limits (inbound and outbound). Cars listed on 'black lists' (no insurance, stolen, etc.) generate an alarm in the dispatching room, so they can be intercepted by a patrol. As of early 2012, 1 million cars per week are automatically checked in this way.

Canada

The police service in Ontario uses automatic licence plate recognition software to nab drivers behind the wheels of vehicles with Ontario number plates.

Denmark

The technique is tested by the Danish police. It will be in permanent use from the end of 2015.

France

180 gantries over major roads have been built throughout the country. These together with a further 250 fixed cameras is to enable a levy of an eco tax on lorries over 3.5 tonnes. The system is currently being opposed and whilst they may be collecting data on vehicles passing the cameras, no eco tax is being charged.

Germany

On 11 March 2008, the Federal Constitutional Court of Germany ruled that some areas of the laws permitting the use of automated number plate recognition systems in Germany violated the right to privacy. More specifically, the court found that the retention of any sort of information (i.e., number plate data) which was not for any pre-destined use (e.g., for use tracking suspected terrorists or for enforcement of speeding laws) was in violation of German law. These systems were provided by Jenoptik Robot GmbH, and called TraffiCapture.

Hungary

Road gantry traffic enforcement and data point on the M7 highway at Érd, Hungary

In 2012 a state consortium was formed among the Hungarian Ministry of Interior, the National Police Headquarters and the Central Commission of Public Administration and Electronic Services with the aim to install and operate a unified intelligent transportation system (*ITS*) with nationwide coverage by the end of 2015. Within the system, 160 portable traffic enforcement and data-gathering units and 365 permanent gantry installations were brought online with ANPR, speed detection, imaging and statistical capabilities. Since all the data points are connected to a centrally located ITS, each member of the consortium is able to separately utilize its range of administrative and enforcement activities, such as remote vehicle registration and insurance verification, speed, lane and traffic light enforcement and wanted or stolen vehicle interception among others.

Several Hungarian auxiliary police units also use a system called Matrix Police in cooperation with the police. It consists of a portable computer equipped with a web camera that scans the stolen car database using automatic number plate recognition. The system is installed on the dashboard of selected patrol vehicles (PDA-based hand-held versions also exist) and is mainly used to control the license plate of parking cars. As the Auxiliary Police do not have the authority to order moving vehicles to stop, if a stolen car is found, the formal police is informed.

Pakistan

The technique is proposed by Lahore Police; if approved it will be functional by 2018.

Saudi Arabia

Vehicle registration plates in Saudi Arabia use white background, but several vehicle types may have a different background. United States diplomatic plates have the letters 'USD', which in Arabic reads 'DSU' when read from right to left in the direction of Arabic script. There are only 17 Arabic letters used on the registration plates. A Challenge for plates recognition in Saudi Arabia is the size of the digits. Some plates use both Eastern Arabic numerals and the 'Western Arabic' equivalents. A research with source code is available for APNR Arabic digits.

Sweden

The technique is tested by the Swedish police at nine different places in Sweden.

Turkey

Several cities have tested—and some have put into service—the "City Security Administration System", i.e., capital Ankara, has debuted KGYS- "Kent Guvenlik Yonetim Sistemi" which consists of a registration plate number recognition system on the main arteries and city exits. The system has been used with two cameras per lane, one for plate recognition, one for speed detection. Now the system has been widened to network all the registration number cameras together, and enforcing average speed over preset distances. Some arteries have 70Kmh limit, and some 50 kmh, and photo evidence with date-time details are posted to registration address if speed violation is detected. As of 2012, the fine for exceeding the speed limit for more than 30% is approximately US$175.

Ukraine

The project of system integration «OLLI Technology» and the Ministry of Internal Affairs of Ukraine Department of State Traffic Inspection (STI) experiments on the introduction of a modern technical complex which is capable to locate stolen cars, drivers deprived of driving licenses and other problem cars in real time. The Ukrainian complex "Video control" working by a principle of video fixing of the car with recognition of license plates with check under data base.

United Kingdom

An ANPR Equipped Vectra of the Greater Manchester Police force

The Home Office states the purpose of automatic number plate recognition in the United Kingdom is to help detect, deter and disrupt criminality including tackling organised crime groups and terrorists. Vehicle movements are recorded by a network of nearly 8000 cameras capturing between 25 and 30 million ANPR 'read' records daily. These records are stored for up to two years in the National ANPR Data Center, which can be accessed, analysed and used as evidence as part of investigations by UK law enforcement agencies.

In 2012, the UK government enacted the Protection of Freedoms Act which includes several provisions related to controlling and restricting the collection, storage, retention, and use of information about individuals. Under this Act, the Home Office published a code of practice in 2013 for the use of surveillance cameras, including ANPR, by government and law enforcement agencies. The aim of the code is to help ensure their use is "characterised as surveillance by consent, and such consent on the part of the community must be informed consent and not assumed by a system operator. Surveillance by consent should be regarded as analogous to policing by consent." In addition, a set a standards were introduced in 2014 for data, infrastructure, and data access and management.

United States

A City of Alexandria police car equipped with mobile ALPR.

ANPR cameras in operation on the Brooklyn Bridge in New York.

In the United States, ANPR systems are more commonly referred to as ALPR (Automatic License Plate Reader/Recognition) technology, due to differences in language (i.e., "number plates" are referred to as "license plates" in American English)

Mobile ANPR use is widespread among US law enforcement agencies at the city, county, state and federal level. According to a 2012 report by the Police Executive Research Forum, approximately 71% of all US police departments use some form of ANPR. Mobile ANPR is becoming a significant component of municipal predictive policing strategies and intelligence gathering, as well as for recovery of stolen vehicles, identification of wanted felons, and revenue collection from individuals who are delinquent on city or state taxes or fines, or monitoring for “Amber Alerts”. With the widespread implementation of this technology, many U.S. states now issue misdemeanor citations of up to $500 when a license plate is identified as expired or on the incorrect vehicle. Successfully recognized plates may be matched against databases including “wanted person”, “protection order”, missing person, gang member, known and suspected terrorist, supervised release, immigration violator, and National Sex Offender lists. In addition to the real-time processing of license plate numbers, ALPR systems in the US collect (and can indefinitely store) data from each license plate capture. Images, dates, times and GPS coordinates can be stockpiled and can help place a suspect at a scene, aid in witness identification, pattern recognition or the tracking of individuals.

The Department of Homeland Security has proposed a federal database to combine all monitoring systems, which was cancelled after privacy complaints. In 1998, a Washington D.C. police lieutenant pleaded guilty to extortion after blackmailing the owners of vehicles parked near a gay bar. In 2015, the Los Angeles Police Department proposed sending letters to the home addresses of all vehicles that enter areas of high prostitution.

An early, private sector mobile ANPR application has been applications for vehicle repossession and recovery), although the application of ANPR by private companies to collect information from privately owned vehicles or collected from private property (for example, driveways) has become an issue of sensitivity and public debate. Other ALPR uses include parking enforcement, and revenue collection from individuals who are delinquent on city or state taxes or fines. The technology is often featured in the reality TV show *Parking Wars* featured on A&E Network. In the show, tow truck drivers and booting teams use the ALPR to find delinquent vehicles with high amounts of unpaid parking fines.

Average-speed Cameras

ANPR is used for speed limit enforcement in Australia, Austria, Belgium, Dubai (UAE), France, Italy, The Netherlands, Spain, and the UK.

This works by tracking vehicles’ travel time between two fixed points, and calculating the average speed. These cameras are claimed to have an advantage over traditional speed cameras in maintaining steady legal speeds over extended distances, rather than encouraging heavy braking on approach to specific camera locations and subsequent acceleration back to illegal speeds.

Italy

In Italian Highways has developed a monitoring system named Tutor covering more than 2500 km (2012). The Tutor system is also able to intercept cars while changing lanes.

The Netherlands

Average speed cameras (*trajectcontrole*) are in place in the Netherlands since 2002. As of July

2009, 12 cameras were operational, mostly in the west of the country and along the A12. Some of these are divided in several "sections" to allow for cars leaving and entering the motorway.

A first experimental system was tested on a short stretch of the A2 in 1997 and was deemed a big success by the police, reducing overspeeding to 0.66%, compared to 5 to 6% when regular speed cameras were used at the same location. The first permanent average speed cameras were installed on the A13 in 2002, shortly after the speed limit was reduced to 80 km/h to limit noise and air pollution in the area. In 2007, average speed cameras resulted in 1.7 million fines for overspeeding out of a total of 9.7 millions. According to the Dutch Attorney General, the average number of violation of the speed limits on motorway sections equipped with average speed cameras is between 1 and 2%, compared to 10 to 15% elsewhere.

United Kingdom

One of the most notable stretches of average speed cameras in the UK is found on the A77 road in Scotland, with 32 miles (51 km) being monitored between Kilmarnock and Girvan. In 2006 it was confirmed that speeding tickets could potentially be avoided from the 'SPECS' cameras by changing lanes and the RAC Foundation feared that people may play "Russian Roulette" changing from one lane to another to lessen their odds of being caught. However, in 2007 the system was upgraded for multi-lane use and in 2008 the manufacturer described the "myth" as "categorically untrue". There exists evidence that implementation of systems such as SPECS has a considerable effect on the volume of drivers travelling at excessive speeds; on the stretch of road mentioned above (A77 Between Glasgow and Ayr) there has been noted a "huge drop" in speeding violations since the introduction of a SPECS system.

Crime Deterrent

Recent innovations have contributed to the adoption of ANPR for perimeter security and access control applications at government facilities. Within the US, "homeland security" efforts to protect against alleged "acts of terrorism" have resulted in adoption of ANPR for sensitive facilities such as embassies, schools, airports, maritime ports, military and federal buildings, law enforcement and government facilities, and transportation centers. ANPR is marketed as able to be implemented through networks of IP based surveillance cameras that perform "double duty" alongside facial recognition, object tracking, and recording systems for the purpose of monitoring suspicious or anomalous behavior, improving access control, and matching against watch lists. ANPR systems are most commonly installed at points of significant sensitivity, ingress or egress. Major US agencies such as the Department of Homeland Security, the Department of Justice, the Department of Transportation and the Department of Defense have purchased ANPR for perimeter security applications. Large networks of ANPR systems are being installed by cities such as Boston, London and New York City to provide citywide protection against acts of terrorism, and to provide support for public gatherings and public spaces.

The Center For Evidence-Based Crime Policy in George Mason University identifies the following randomized controlled trials of automatic number plate recognition technology as very rigorous.

Authors	Study	Results
Braga, A. A., & Bond, B. J.	"*Policing crime and disorder hot spots: A randomized, controlled trial*", 2008	Declines for disorder calls for service in target hot spots.
Hegarty, T., Williams, L. S., Stanton, S., & Chernoff, W.	"*Evidence-Based Policing at Work in Smaller Jurisdictions*", 2014	Decrease in crimes and calls for service across all hot spots during the trial. No statistically significant difference in crimes found between the visibility and visibility-activity hot spots.

Enterprise Security and Services

In addition to government facilities, many private sector industries with facility security concerns are beginning to implement ANPR solutions. Examples include casinos, hospitals, museums, parking facilities, and resorts. In the US, private facilities typically cannot access government or police watch lists, but may develop and match against their own databases for customers, VIPs, critical personnel or "banned person" lists. In addition to providing perimeter security, private ANPR has service applications for valet / recognized customer and VIP recognition, logistics and key personnel tracking, sales and advertising, parking management, and logistics (vendor and support vehicle tracking).

Traffic Control

Video Tolling at Schönberg, Austria

Many cities and districts have developed traffic control systems to help monitor the movement and flow of vehicles around the road network. This had typically involved looking at historical data, estimates, observations and statistics, such as:

- Car park usage
- Pedestrian crossing usage
- Number of vehicles along a road
- Areas of low and high congestion
- Frequency, location and cause of road works

CCTV cameras can be used to help traffic control centres by giving them live data, allowing for traffic management decisions to be made in real-time. By using ANPR on this footage it is possible to monitor the travel of individual vehicles, automatically providing information about the speed and flow of various routes. These details can highlight problem areas as and when they occur and help the centre to make informed incident management decisions.

Some counties of the United Kingdom have worked with Siemens Traffic to develop traffic monitoring systems for their own control centres and for the public. Projects such as Hampshire County Council's ROMANSE provide an interactive and real-time website showing details about traffic in the city. The site shows information about car parks, ongoing road works, special events and footage taken from CCTV cameras. ANPR systems can be used to provide average point-to-point journey times along particular routes, which can be displayed on a variable-message sign(VMS) giving drivers the ability to plan their route. ROMANSE also allows travellers to see the current situation using a mobile device with an Internet connection (such as WAP, GPRS or 3G), allowing them to view mobile device CCTV images within the Hampshire road network.

The UK company Trafficmaster has used ANPR since 1998 to estimate average traffic speeds on non-motorway roads without the results being skewed by local fluctuations caused by traffic lights and similar. The company now operates a network of over 4000 ANPR cameras, but claims that only the four most central digits are identified, and no numberplate data is retained.

IEEE Intelligent Transportation Systems Society published some papers on the plate number recognition technologies and applications.

Electronic Toll Collection

Toll Roads

The FasTrak system in Orange County, California, uses ANPR and radio transponders

Film showing the approach to and passing of a toll station in Italy, using a Telepass OBU. Note the yellow Telepass lane signs and road markings and the sound emitted by the OBU when passing the lane

Ontario's 407 ETR highway uses a combination of ANPR and radio transponders to toll vehicles entering and exiting the road. Radio antennas are located at each junction and detect the transponders, logging the unique identity of each vehicle in much the same way as the ANPR system does. Without ANPR as a second system it would not be possible to monitor all the traffic. Drivers who opt to rent a transponder for C$2.55 per month are not charged the "Video Toll Charge" of C$3.60 for using the road, with heavy vehicles (those with a gross weight of over 5,000 kg) being required to use one. Using either system, users of the highway are notified of the usage charges by post.

There are numerous other electronic toll collection networks which use this combination of Radio frequency identification and ANPR. These include:

- The Golden Gate Bridge in San Francisco, California, began using an all-electronic tolling system combining Fastrak and ANPR on March 27, 2013.

- NC Quick Pass for the Interstate 540 (North Carolina) Triangle Expressway in Wake County, North Carolina
- Bridge Pass for the Saint John Harbour Bridge in Saint John, New Brunswick
- Quickpass at the Golden Ears Bridge, crossing the Fraser River between Langley and Maple Ridge
- e-TAG, Australia
- FasTrak in California, United States
- Highway 6 in Israel
- Tunnels in Hong Kong
- Autopista Central in Santiago, Chile (site in Spanish)
- E-ZPass in New York, New Jersey, Pennsylvania, Massachusetts (as Fast Lane until 2012), Virginia (formerly Smart Tag), and other states. Maryland Route 200 uses a combination of E-ZPass and ANPR.
- TollTag in North Texas and EZ-Tag in Houston, Texas.
- I-Pass in Illinois
- Pike Pass in Oklahoma
- Peach Pass I-85 Atlanta, GA Gwinnett County
- OGS (Otomatik Geçiş Sistemi) used at Bosphorus Bridge, Fatih Sultan Mehmet Bridge, and Trans European Motorway entry points in İstanbul, Turkey
- M50 Westlink Toll in Dublin, Ireland
- Hi-pass in South Korea
- Northern Gateway, SH 1, Auckland, New Zealand
- Evergreen Point Floating Bridge, Seattle, and Washington State Route 167 HOT-lanes in western Washington
- ETC in Taiwan

* Sun-pass In Florida

Portugal

Portuguese roads have old highways with toll station where drivers can pay with cards and also lanes where there are electronic collection systems. However most new highways only have the option of electronic toll collection system. The electronic toll collection system comprises three different structures:

- ANPR which works with infrared cameras and reads license plates from every vehicle
- Lasers for volumetric measurement of the vehicle to confirm whether it is a regular car or an SUV or truck, as charges differ according to the type of vehicle
- RFID-like to read on-board smart tags.

When the smart tag is installed in the vehicle, the car is quickly identified and owners bank account is automatically deducted. This process is realized at any speed up to over 250 km per hour. If the car does not have the smart tag, the driver is required to go to a pay station to pay the tolls between 3rd and 5th day after with a surplus charge. If he fails to do so, the owner is sent a letter home with a heavy fine. If this is not paid, it increases five-fold and after that, the car is inserted into a police database for vehicle impounding. This system is also used in some limited access areas of main cities to allow only entry from pre-registered residents. It is planned to be implemented both in more roads and in city entrance toll collection/access restriction. The efficacy of the system is considered to be so high that it is almost impossible for the driver to complain.

London Congestion Charge

The London congestion charge scheme uses 230 cameras and ANPR to help monitor vehicles in the charging zone

The London congestion charge is an example of a system that charges motorists entering a payment area. Transport for London (TfL) uses ANPR systems and charges motorists a daily fee of £10 paid before 10pm if they enter, leave or move around within the congestion charge zone between 7 a.m. and 6:00 p.m., Monday to Friday. A reduced fee of £9 is paid by vehicle owners who sign up for the automatic deduction scheme. Fines for traveling within the zone without paying the charge are £60 per infraction if paid before the deadline, doubling to £120 per infraction thereafter.

There are currently 1,500 cameras which use automatic number plate recognition (ANPR) technology. There are also a number of mobile camera units which may be deployed anywhere in the zone.

It is estimated that around 98% of vehicles moving within the zone are caught on camera. The video streams are transmitted to a data centre located in central London where the ANPR software deduces the registration plate of the vehicle. A second data centre provides a backup location for image data.

Both front and back number plates are being captured, on vehicles going both in and out – this gives up to four chances to capture the number plates of a vehicle entering and exiting the zone. This list is then compared with a list of cars whose owners/operators have paid to enter the zone – those that have not paid are fined. The registered owner of such a vehicle is looked up in a database provided by the DVLA.

Sweden

In Stockholm, Sweden, ANPR is used for the Stockholm congestion tax, owners of cars driving into or out of the inner city must pay a charge, depending on the time of the day. From 2013, also for the Gothenburg congestion tax, which also includes vehicles passing the city on the main highways.

Private use

Several UK companies and agencies use ANPR systems. These include Vehicle and Operator Services Agency (VOSA), Driver and Vehicle Licensing Agency (DVLA) and Transport for London.

Other uses

ANPR systems may also be used for/by:

- Section control, to measure average vehicle speed over longer distances.
- Border crossings
- Automobile repossessions
- Petrol stations to log when a motorist drives away without paying for their fuel.
- A marketing tool to log patterns of use
- Targeted advertising, a-la "Minority Report"-style billboards.
- Traffic management systems, which determine traffic flow using the time it takes vehicles to pass two ANPR sites
- Analyses of travel behaviour (route choice, origin-destination etc.) for transport planning purposes
- Drive Through Customer Recognition, to automatically recognize customers based on their license plate and offer them the items they ordered the last time they used the service.
- To assist visitor management systems in recognizing guest vehicles.
- Police and Auxiliary Police
- Car parking companies
- Hotels

Challenges

Circumvention

Vehicle owners have used a variety of techniques in an attempt to evade ANPR systems and road-rule enforcement cameras in general. One method increases the reflective properties of the lettering and makes it more likely that the system will be unable to locate the plate or produce a high enough level of contrast to be able to read it. This is typically done by using a plate cover or a spray, though claims regarding the effectiveness of the latter are disputed. In most jurisdictions, the covers are illegal and covered under existing laws, while in most countries there is no law to disallow the use of the sprays. Other users have attempted to smear their license plate with dirt or utilize covers to mask the plate.

Novelty frames around Texas license plates were made illegal in Texas on 1 September 2003 by Texas Senate Bill 439 because they caused problems with ANPR devices. That law made it a Class C misdemeanor (punishable by a fine of up to US $200), or Class B (punishable by a fine of up to US $2,000 and 180 days in jail) if it can be proven that the owner did it to deliberately obscure their plates. The law was later clarified in 2007 to allow Novelty frames.

If an ANPR system cannot read the plate, it can flag the image for attention, with the human operators looking to see if they are able to identify the alphanumerics.

In order to avoid surveillance or penalty charges, there has been an upsurge in car cloning. This is usually achieved by copying registration plates from another car of a similar model and age. This can be difficult to detect, especially as cloners may change the registration plates and travel behavior to hinder investigations.

In 2013 researchers at Sunflex Zone Ltd created a privacy license plate frame that uses near infrared light to make the license plate unreadable to license plate recognition systems.

Controversy

The introduction of ANPR systems has led to fears of misidentification and the furthering of *1984*-style surveillance. In the United States, some such as Gregg Easterbrook oppose what they call "machines that issue speeding tickets and red-light tickets" as the beginning of a slippery slope towards an automated justice system:

> "A machine classifies a person as an offender, and you can't confront your accuser because there is no accuser... can it be wise to establish a principle that when a machine says you did something illegal, you are presumed guilty?"

Similar criticisms have been raised in other countries. Easterbrook also argues that this technology is employed to maximize revenue for the state, rather than to promote safety. The electronic surveillance system produces tickets which in the US are often in excess of $100, and are virtually impossible for a citizen to contest in court without the help of an attorney. The revenues generated by these machines are shared generously with the private corporation that builds and operates them, creating a strong incentive to tweak the system to generate as many tickets as possible.

Older systems had been notably unreliable; in the UK this has been known to lead to charges being made incorrectly with the vehicle owner having to pay £10 in order to be issued with proof (or not) of the offense. Improvements in technology have drastically decreased error rates, but false accusations are still frequent enough to be a problem.

Perhaps the best known incident involving the abuse of an ANPR database in North America is the case of *Edmonton Sun* reporter Kerry Diotte in 2004. Diotte wrote an article critical of Edmonton police use of traffic cameras for revenue enhancement, and in retaliation was added to an ANPR database of "high-risk drivers" in an attempt to monitor his habits and create an opportunity to arrest him. The police chief and several officers were fired as a result, and The Office of the Privacy Commissioner of Canada expressed public concern over the "growing police use of technology to spy on motorists."

Other concerns include the storage of information that could be used to identify people and store details about their driving habits and daily life, contravening the Data Protection Act along with similar legislation. The laws in the UK are strict for any system that uses CCTV footage and can identify individuals.

Also of concern is the safety of the data once it is mined, following the discovery of police surveillance records lost in a gutter.

There is also a case in the UK for saying that use of ANPR cameras is against the law under the Regulation of Investigatory Powers Act 2000. The breach exists, some say, in the fact that ANPR is used to monitor the activities of law-abiding citizens and treats everyone like the suspected criminals intended to be surveyed under the act. The police themselves have been known to refer to the system of ANPR as a "24/7 traffic movement database" which is a diversion from its intended purpose of identifying vehicles involved in criminal activities. The opposing viewpoint is that where the plates have been cloned, a 'read' of an innocent motorist's vehicle will allow the elimination of that vehicle from an investigation by visual examination of the images stored. Likewise, stolen vehicles are read by ANPR systems between the time of theft and report to the Police, assisting in the investigation.

The *Associated Press* reported in August 2011 that New York Police Department cars and license plate tracking equipment purchased with federal HIDTA (High Intensity Drug Trafficking Area) funds were used to spy on Muslims at mosques, and to track the license plate numbers of worshippers. Police in unmarked cars outfitted with electronic license plate readers would drive down the street and automatically catalog the plates of everyone parked near the mosque, amassing a covert database that would be distributed among officers and used to profile Muslims in public.

In 2013 the American Civil Liberties Union released 26,000 pages of data about ANPR systems obtained from local, state, and federal agencies through freedom of information laws. "The documents paint a startling picture of a technology deployed with too few rules that is becoming a tool for mass routine location tracking and surveillance" wrote the ACLU. The ACLU reported that in many locations the devices were being used to store location information on vehicles which were not suspected of any particular offense. "Private companies are also using license plate readers and sharing the information they collect with police with little or no oversight or privacy protections. A lack of regulation means that policies governing how long our location data is kept vary widely," the ACLU said. In 2012 the ACLU filed suit against the Department of Homeland Security, which funds many local and state ANPR programs through grants, after the agency failed to provide access to records the ACLU had requested under the Freedom of Information Act about the programs.

Plate Inconsistency and Jurisdictional Differences

Many ANPR systems claim accuracy when trained to match plates from a single jurisdiction or region, but can fail when trying to recognize plates from other jurisdictions due to variations in format, font, color, layout, and other plate features. Some jurisdictions offer vanity or affinity plates (particularly in the US), which can create many variations within a single jurisdiction.

From time to time, US states will make significant changes in their license plate protocol that will affect OCR accuracy. They may add a character or add a new license plate design. ALPR systems must adapt to these changes quickly in order to be effective. Another challenge with ALPR systems is that some states have the same license plate protocol. For example, more than one state uses the standard three letters followed by four numbers. So each time the ALPR systems alarms, it is the user's responsibility to make sure that the plate which caused the alarm matches the state associated with the license plate listed on the in-car computer. For maximum effectiveness, an ANPR system should be able to recognize plates from any jurisdiction, and the jurisdiction to which they are associated, but these many variables make such tasks difficult.

Currently at least one US ANPR provider (PlateSmart) claims their system has been independently reviewed as able to accurately recognize the US state jurisdiction of license plates, and one European ANPR provider claims their system can differentiate all EU plate jurisdictions.

Accuracy and Measurement of ANPR System Performance

A 2008 article in *Parking Trend International* discussed a disparity in claimed vs. experienced license plate recognition read rates, with manufacturers claiming that their recognition engines can correctly report 98% of the time, although customers experience only 90% to 94% success, even with new equipment under perfect conditions. Early systems were reportedly only 60% to 80% reliable.

True system error rate is the product of its subsystem error rates (image capture, license plate image extraction, LP image interpretation); slight increases in subsystem error rates can produce dramatic reductions of read rates. The effects of real-world interfering factors on read rate are not uniformly specified or tested by manufacturers. The article states "there is a need for the industry to adopt a standard performance measurement protocol to enable potential customers assess the best fit for their particular requirements."

Traffic Enforcement Camera

A traffic enforcement camera (also red light camera, road safety camera, road rule camera, photo radar, photo enforcement, speed camera, Gatso, safety camera, bus lane camera, flash for cash, Safe-T-Cam, depending on use) is a camera which may be mounted beside or over a road or installed in an enforcement vehicle to detect traffic regulation violations, including speeding, vehicles going through a red traffic light, unauthorized use of a bus lane, or for recording vehicles inside a congestion charge area. It may be linked to an automated ticketing system.

The latest automatic number plate recognition systems can be used for the detection of average speeds and raise concerns over loss of privacy and the potential for governments to establish mass surveillance of vehicle movements and therefore by association also the movement of the vehicle's owner. Vehicles owners are often required by law to identify the driver of the vehicle and a case was taken to the European Court of Human Rights which found that human rights were not being breached. Some groups, such as the National Motorists Association in the USA, claim that systems "encourage ... revenue-driven enforcement" rather than the declared objectives.

Types

Automatic speed enforcement gantry or "*Lombada Eletrônica*" with ground sensors at Brasilia, D.F.

Gatso Mobile Speed Camera, used in Victoria, Australia. The camera is mounted on the passenger side dash, whilst the black box on the front is the radar unit.

Bus Lane Enforcement

Some bus lane enforcement cameras use a sensor in the road, which triggers a number plate recognition camera which compares the vehicle registration plate with a list of approved vehicles and records images of other vehicles. Other systems use a camera mounted on the bus, for example in London where they monitor Red routes on which stopping is not allowed for any purpose (other than taxis and disabled parking permit holders).

On Monday, February 23, 2009, New York City announced testing camera enforcement of bus lanes on 34th Street in Midtown Manhattan where a New York City taxi illegally using the bus lanes would face a fine of $150 adjudicated by the New York City Taxi and Limousine Commission.

In October 2013, in Melbourne (Australia), Melbourne Airport introduce 7 Automatic Number Plate Recognition (ANPR) cameras in their bus forecourt to monitor bus lanes and provide charging points based on vehicle type and the dwell time of each vehicle. Entry and Exit cameras determine the length of stay and provide alerts for unregistered or vehicles of concern via onscreen, email or SMS based alerts. This system was the first of several Sensor Dynamics based ANPR solutions.

Melbourne Airport were the first Australian Airport to use ANPR technology to charge buses for access bus pick up lanes.

Red Light Enforcement

Redflex red light camera in Springfield, Ohio, USA.

A red light camera is a traffic camera that takes an image of a vehicle that goes through an intersection where the light is red. The system continuously monitors the traffic signal and the camera is triggered by any vehicle entering the intersection above a preset minimum speed and following a specified time after the signal has turned red.

Speed Limit Enforcement

Speed enforcement cameras are used to monitor compliance with speed limits, which may use Doppler radar, LIDAR or automatic number plate recognition. Other speed enforcement systems are also used which are not camera based.

Fixed or mobile speed camera systems that measure the time taken by a vehicle to travel between two or more fairly distant sites (from several hundred metres to several hundred kilometres apart) are called automatic number plate recognition (ANPR) cameras. These cameras time vehicles over a known fixed distance, and then calculate the vehicle's average speed for the journey. The name derives from the fact that the technology uses infrared cameras linked to a computer to read a vehicle's registration number and identify it in real-time.

Stop Sign Enforcement

In 2007, the Mountains Recreation and Conservation Authority (MRCA), in California, installed the first stop sign cameras in the United States. The five cameras are located in state parks such as Franklin Canyon Park and Temescal Gateway Park. The operator, Redflex Traffic Systems Inc., is paid $20 per ticket. The fine listed on the citation is $100. In 2010, a class action lawsuit was filed against MRCA.

Number Plate Recognition Systems

Automatic number plate recognition can be used for purposes unrelated to enforcement of traffic rules. In principle any agency or person with access to data either from traffic cameras or cameras installed for other purposes can track the movement of vehicles for any purpose.

In Australia's SAFE-T-CAM system, ANPR technology is used to monitor long distance truck drivers to detect avoidance of legally prescribed driver rest periods.

The United Kingdom's police ANPR system logs all the vehicles passing particular points in the national road network, allowing authorities to track the movement of vehicles and individuals across the country.

In the UK an 80-year-old pensioner John Catt and his daughter Linda were stopped by City of London Police while driving in London, UK in 2005. They had their vehicle searched under section 44 of the Terrorism Act 2000 and were threatened with arrest if they refused to answer questions. After they complained formally, it was discovered they were stopped when their car was picked up by roadside ANPR CCTV cameras; it had been flagged in the Police National Computer database when they were seen near EDO MBM demonstrations in Brighton. Critics point out that the Catts had been suspected of no crime, however the UK's mass surveillance infrastructure allowed them to be targeted due to their association.

Other

- Congestion charge cameras to detect vehicles inside the chargeable area which have not paid the appropriate fee
- High-occupancy vehicle lane cameras to identify vehicles violating occupancy requirements.
- Level crossing cameras to identifying vehicles crossing railways at grade
- Noise pollution cameras that record evidence of heavy vehicles that break noise regulations by using compression release engine brakes
- Parking cameras which issue citations to vehicles which are illegally parked or which were

not moved from a street at posted times.

- Toll-booth cameras to identify vehicles proceeding through a toll booth without paying the toll
- Turn cameras at intersections where specific turns are prohibited on red. This type of camera is mostly used in cities or heavy populated areas.
- Automatic number plate recognition systems can be used for multiple purposes, including identifying untaxed and uninsured vehicles, stolen cars and potentially mass surveillance of motorists .
- Bus lane cameras that detect vehicles that should not be in the bus lane. These may be mounted on buses themselves as well as by the roadside.

Fixed camera systems can be housed in boxes, or mounted on poles beside the road, or attached to gantries over the road, or to overpasses or bridges. Cameras can be concealed, for example in garbage bins.

Mobile speed cameras may be hand-held, tripod mounted, or vehicle-mounted. In vehicle-mounted systems, detection equipment and cameras can be mounted to the vehicle itself, or simply tripod mounted inside the vehicle and deployed out a window or door. If the camera is fixed to the vehicle, the enforcement vehicle does not necessarily have to be stationary, and can be moved either with or against the flow of traffic. In the latter case, depending on the direction of travel, the target vehicle's relative speed is either added or subtracted from the enforcement vehicle's own speed to obtain its actual speed. The speedometer of the camera vehicle needs to be accurately calibrated.

Some number plate recognition systems can also be used from vehicles.

Controversy

Legal Issues

Various legal issues arise from such cameras and the laws involved in how cameras can be placed and what evidence is necessary to prosecute a driver varies considerably in different legal systems.

One issue is the potential conflict of interest when private contractors are paid a commission based on the number of tickets they are able to issue. Pictures from the San Diego red light camera systems were ruled inadmissible as court evidence in September 2001. The judge said that the "total lack of oversight" and "method of compensation" made evidence from the cameras "so untrustworthy and unreliable that it should not be admitted".

Some U.S. states and provinces of Canada, such as Alberta operate "owner liability", where it is the registered owner of the vehicle who must pay all such fines, regardless who was driving the vehicle at the time of the offense, although they do release the owner from liability if he or she identifies the actual driver and that person pays the fine.

In a few U.S. states (including California), the cameras are set up to get a "face photo" of the driver. This has been done because in those states red light camera tickets are criminal violations, and criminal charges must always name the actual violator. In California, that need to identify the actual violator has led to the creation of a unique investigatory tool, the fake "ticket."

In April, 2000, two motorists who were caught speeding in the United Kingdom challenged the *Road Traffic Act 1988*, which required the keeper of a car to identify the driver at a particular time as being in contradiction to the Human Rights Act 1998 on the grounds that it amounted to a 'compulsory confession', also that since the camera partnerships included the police, local authorities, Magistrates Courts Service (MCS) and Crown Prosecution Service (CPS) which had a financial interest in the fine revenue that they would not get a fair trial. Their plea was initially granted by a judge then overturned but was the heard by the European Court of Human Rights (ECtHR), and the European Court of Justice (ECJ). In 2007 the European Court of Human Rights found there was no breach of article 6 in requiring the keepers of cars caught speeding on camera to provide the name of the driver.

Accuracy

In December, 2012, Speed Camera Contractor Xerox Corporation admitted that cameras they had deployed in Baltimore city were producing erroneous speed readings, and that 1 out of every 20 citations issued at some locations were due to errors. The erroneous citations included at least one issued to a completely stationary car, a fact revealed by a recorded video of the alleged violation.

In the city of Fort Dodge, Iowa, speed camera contractor Redspeed discovered a location where drivers of school buses, big panel trucks and similar vehicles have been clocked speeding by the city's mobile speed camera and radar unit even though they were obeying the 25 mph speed limit. The errors were due to what was described as an "electromagnetic anomaly".

Where verification photos are recorded on a time sequence, allowing the calculation of actual speed, these have been used to challenge the accuracy of speed cameras in court. Motorists in Prince George's County, Maryland, have successfully challenged tickets from Optotraffic speed cameras where they were incorrectly ticketed at over 15 mph over the limit. However, Prince George County no longer allows time-distance calculations as a defense in cases where "the equipment was calibrated and validated, or is self calibrating". The National Highway Traffic Safety Administration standards for "across the road radar" state that "If the ATR device is to be considered for unattended operation, the manufacturer shall provide a secondary method for verifying that the evidential recorded image properly identifies the target vehicle and reflects this vehicle's true speed, as described in §5.18.2. This may be accomplished by means of a second, appropriately delayed image showing the target vehicle crossing a specified reference line."

Surveillance

- Police and government have been accused of "Big Brother tactics" in over-monitoring of public roads, and of "revenue raising" in applying cameras in deceptive ways to increase government revenue rather than improve road safety.

Revenue, Not Safety

- In 2010, a campaign was set up against a speed camera on a dual carriageway in Poole, Dorset in a 30 mph area in the United Kingdom, which had generated £1.3m of fines every year since 1999. The initial Freedom of information request was refused and the information was only released after an appeal to the Information Commissioner.

- In May, 2010, the new Coalition government said that the 'Labour's 13-year war on the motorist is over' and that the new government 'pledged to scrap public funding for speed cameras' In July Mike Penning, the Road safety minister reduced the Road Safety Grant for the current year to Local Authorities from £95 million to £57 million, saying that local authorities had relied too heavily on safety cameras for far too long and that he was pleased that some councils were now focusing on other road safety measures. It is estimated that as a result the Treasury is now distributing £40 million less in Road Safety Grant than is raised from fines in the year. Dorset and Essex announced plans to review camera provision with a view to possibly ending the scheme in their counties, however Dorset strongly affirmed its support for the scheme, albeit reducing financial contributions in line with the reduction in government grant. Seven counties also announced plans to turn off some or all of their cameras, amidst warnings from the country's most senior traffic policeman that this would result in an increase in deaths and injuries. Gloucestershire cancelled plans to update cameras and has reduced or cancelled maintenance contracts.
- In August 2010, the Oxfordshire, UK speed cameras had been switched off because of lack of finance due to government funding policy changes. The cameras were switched back on in April 2011 after a new source of funding was found for them. Following rule changes on the threshold for offering "Speed Awareness Courses" as an alternative to a fine and licence points for drivers, and given that the compulsory fees charged for such courses go directly to the partnerships rather than directly to central government as is the case for fine revenues, the partnership will be able to fund its operations from course fees. Compared with the same period in the previous year with the cameras still switched on, the number of serious injuries that occurred during the same period with the cameras switched off was exactly the same - at 13 - and the number of slight injuries was 15 more at 70, resulting from 62 crashes - 2 more than when the cameras were still operating. There were no fatalities during either period.

Unpopularity

Claims of popular support are disputed by elections in the US, where the camera companies often sue to keep it off the ballot, and camera enforcement often loses by a wide margin. Automated enforcement is opposed by some motorists and motoring organizations as strictly for revenue generating. They have also been rejected in some places by referendum.

- The first speed camera systems in the USA was in Friendswood, Texas in 1986 and La Marque, Texas in 1987. Neither program lasted more than a few months before public pressure forced them to be dropped.
- In 1991, cameras were rejected in referendum in Peoria, Arizona; voters were the first to reject cameras by a 2-1 margin.
- In 1992, cameras were rejected by voters in referenda in Batavia, Illinois.
- Anchorage, Alaska rejected cameras in a 1997 referendum.

- In 2002, the state of Hawaii experimented with speed limit enforcement vans but they were withdrawn months later due to public outcry.
- A 2002 Australian survey found that "The community generally believes that enforcement intensities should either stay the same or increase", with 40% of those surveyed saying that they thought that the number of speed cameras on the road should be increased, 43% saying that they thought the number should stay the same, and 13% saying that they thought that the number should be decreased.
- In 2005, the Virginia legislature declined to reauthorize its red light camera enforcement law after a study questioned their effectiveness, only to reverse itself in 2007 and allow cameras to return to any city with a population greater than 10,000.
- Steubenville, Ohio rejected cameras in a 2006 referendum.
- A 2007 literature review of the benefits and barriers to implementation of automated speed enforcement in the U.S. stated that "In general, the results of [public opinion] surveys indicate that a majority of respondents support automated enforcement. However, the margins of support vary widely, from a low of 51 percent in Washington, D.C. to a high of 77 percent in Scottsdale, Arizona."
- In 2009, a petition was started in the town of College Station, Texas which requested that all red light cameras be dismantled and removed from all of the town's intersections. Enough signatures were captured to put the measure on the November 2009 general election ballot. After an extensive battle between the College Station city council and the opposing sides, both for and against red light cameras, the voters voted to eliminate the red light cameras throughout the entire city. By the end of November the red light cameras were taken down.
- On May 4, 2010, an ordinance authorizing the use of speed cameras in the town of Sykesville, Maryland was put to a referendum, in which 321 out of 529 voters (60.4%) voted against the cameras. The turnout for this vote was greater than the number of voters in the previous local Sykesville election for mayor where 523 voted.
- Arizona decided not to renew their contract with Redflex in 2011 following a study of their statewide 76 photo enforcement cameras. Reasons given included less than expected revenue due to improved compliance, mixed public acceptance and mixed accident data.

Effectiveness

- Aside from the issues of legality in some countries and states and of sometime opposition the effectiveness of speed cameras is very well documented. The introduction to The Effectiveness of Speed Cameras A review of evidence by Richard Allsop includes the following in the foreword by Professor Stephen Glaister, Director of the RAC (Royal Automobile Club). "While this report fully lays out the background to the introduction of speed cameras and the need for speed limits, its job is not to justify why the national limits are what they are; a review of speed limits to see whether they are soundly based is for another day. What it has done is to show that at camera sites, speeds have been reduced, and that as a result, collisions resulting in injuries have fallen. The government has said that a decision on whether speed cameras should be funded must be taken at a local level. With the current pressure on public funds, there will be – indeed there already are – those who say that what little money there is can be better spent. This report begs to differ. The devices are already there; they demonstrate value for money, yet are not significant revenue raisers for the Treasury; they are shown to save lives; and despite the headlines, most people accept the need for them. Speed cameras should never be the only weapon in the road safety armoury, but neither should they be absent from the battle."

- The town of Swindon abandoned the use of fixed cameras in 2009, questioning their cost effectiveness with the cameras being replaced by vehicle activated warning signs and enforcement by police using mobile speed cameras: in the nine months following the switch-off there was a small reduction in accident rates which had changed slightly in similar periods before and after the switch off (Before: 1 fatal, 1 serious and 13 slight accidents. Afterwards: no fatalities, 2 serious and 12 slight accidents). The journalist George Monbiot claimed that the results were not statistically significant highlighting earlier findings across the whole of Wiltshire that there had been a 33% reduction in the number of people killed and seriously injured generally and a 68% reduction at camera sites during the previous 3 years.
- The 2010 Cochrane Review of speed cameras for the prevention of road traffic injuries and deaths reported that all 28 studies accepted by the authors found the effect of speed cameras to be a reduction in all crashes, injury crashes, and death or severe injury crashes. "Twenty eight studies measured the effect on crashes. All 28 studies found a lower number of crashes in the speed camera areas after implementation of the program. In the vicinity of camera sites, the reductions ranged from 8% to 49% for all crashes, with reductions for most studies in the 14% to 25% range. For injury crashes the decrease ranged between 8% to 50% and for crashes resulting in fatalities or serious injuries the reductions were in the range of 11% to 44%. Effects over wider areas showed reductions for all crashes ranging from 9% to 35%, with most studies reporting reductions in the 11% to 27% range. For crashes resulting in death or serious injury reductions ranged from 17% to 58%, with most studies reporting this result in the 30% to 40% reduction range. The studies of longer duration showed that these positive trends were either maintained or improved with time."
- In January 2011 Edmonton, Alberta cancelled all 100,000 "Speed On Green" tickets issued in the previous 14 months due to concerns about camera reliability.
- According to the 2003 NCHRP Synthesis 310, "RLR automated enforcement can be an effective safety countermeasure....[I]t appears from the findings of several studies that, in general, RLR cameras can bring about a reduction in the more severe angle crashes with, at worst, a slight increase in less severe rear-end crashes. However it noted that "there is not enough empirical evidence based on proper experimental design procedures to state this conclusively."
- The 2010 report, "The Effectiveness of Speed Cameras A review of evidence", by Richard Allsop concludes "The findings of this review for the RAC Foundation, though reached independently, are essentially consistent with the Cochrane Review conclusions. They are also broadly consistent with the findings of a meta-analysis reported in the respected

Handbook of Road Safety Measures, of 16 studies, not including the four-year evaluation report, of the effects of fixed cameras on numbers of collisions and casualties."

- A recent study conducted in Alabama reveals that RLCs seem to have a slight impact on the clearance lost time; the intersections equipped with RLCs are half a second less in use compared with those without cameras; and highway capacity manual estimates a shorter lost time and thus may overestimate the intersection's capacity.
- In the UK the town of Swindon abandoned the use of fixed speed cameras, arguing that the cost did not represent an effective way to reduce road accidents. Within four years the town was the safest town to drive in the UK, based on accident rates per 1,000 registered vehicles: a result linked by the Local Authority Member for Council Transformation, Transport and Strategic Planning to the removal of speed cameras and resultant additional funding for road safety, alongside close working with the police.

Critical Response

Online websites, like Photo Radar Scam and BantheCams.org, have been created in reaction to the rising use of traffic cameras. Their primary goal, as stated by BantheCams.org, is to "educate and equip local citizens with a way to combat the abuse of power now being exercised by local and state governments with the regards to the use of electronic surveillance devices."

Groups like NHTSA (National Highway Traffic Safety Administration) have encouraged the usage of automated speed enforcement to help improve general road safety and to decrease crash rates.

Avoidance/Evasion

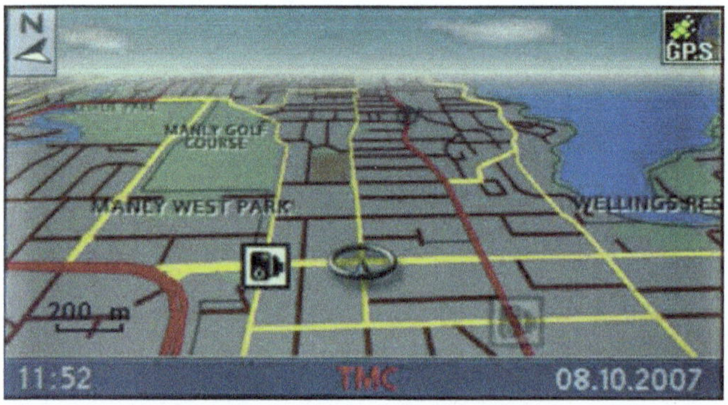

A GPS map showing speed camera POI information overlaid onto it

To avoid detection or prosecution, drivers may:

- Drive at or below the legal speed. Using apps like Live Speed Limit Alerts can help with audible speed limit alerts.
- Brake just before a camera in order to travel past its sensor below the speed limit. This is, however, a cause of collisions. Or brake suddenly, which results in rear-end crashes.
- Use GPS navigation devices which contain databases of known camera locations to alert them in advance. These databases may, in some cases, be updated in near-realtime. The

use of GPS devices to locate speed cameras is illegal in some jurisdictions, such as France. In Australia, the use of GPS devices within the category of intelligent speed adaptation are being encouraged.

- Although ineffective, passive laser detectors or radar detectors that detect when the vehicle's speed is being monitored exist. Use of these devices may be illegal in some jurisdictions such as France.
- Install active laser jammer or radar jammer devices which actively transmit signals that interfere with the measuring device. These devices are illegal in many jurisdictions.
- Remove, falsify, obscure or modify vehicle license plate. Tampering with number plates or misrepresenting them is illegal in most jurisdictions.
- Vandalize the camera. This is illegal in all jurisdictions.

In August, 2010, a fast driving Swedish driver reportedly avoided several older model speed cameras, but was detected by a new model, as traveling at 300 km/h (186 mph), resulting in the world's second largest speeding fine to date.

In the past, it was possible to avoid detection by changing lanes when SPECS average speed cameras were in use as they measured a vehicle's speed over distance in one lane only. Since 2007, measures were taken to mitigate this limitation. Although the cameras do operate in pairs on single lanes (it is a limitation of the technology not a restriction in the type approval) the authorities now install the cameras such that the monitored length of road overlaps between multiple camera pairs. The driver cannot tell which cameras are 'entry' and which are 'exit' making it difficult to know when to change lane.

History

Device for speed control in the Hague, newsreel from October 1940

Older traffic enforcement camera in Ludwigsburg, Germany

The concept of the speed camera can be dated back to at least 1905; Popular Mechanics reports on a patent for a "Time Recording Camera for Trapping Motorists" that enabled the operator to take time-stamped images of a vehicle moving across the start and endpoints of a measured section of road. The timestamps enabled the speed to be calculated, and the photo enabled identification of the driver.

The Dutch company *Gatsometer BV*, which was founded in 1958 by rally driver Maurice Gatsonides, produced the 'Gatsometer'. Gatsonides wished to better monitor his average speed on a race track and invented the device in order to improve his lap times. The company later started supplying these devices as police speed enforcement tools. The first systems introduced in the late 1960s used film cameras to take their pictures. Gatsometer introduced the first red light camera in 1965, the first radar for use with road traffic in 1971 and the first mobile speed traffic camera in 1982;

From the late 1990s, digital cameras began to be introduced. Digital cameras can be fitted with a network connection to transfer images to a central processing location automatically, so they have advantages over film cameras in speed of issuing fines, maintenance and operational monitoring. However, film-based systems may provide superior image quality in the variety of lighting conditions encountered on roads, and are required by courts in some jurisdictions. New film-based systems are still being sold, but digital pictures are providing greater versatility and lower maintenance and are now more popular with law enforcement agencies.

Gallery

A red-light and speed camera in Darwin, Northern Territory, Australia

Dazzle camouflaged speed camera as an art project in Loipersdorf, Austria

A red-light camera in use in Beaverton, Oregon, USA

Parking Guidance and Information

Parking guidance and information (PGI) systems, or car park guidance systems, present drivers with dynamic information on parking within controlled areas. The systems combine traffic monitoring, communication, processing and variable message sign technologies to provide the service.

Modern parking lots utilize a variety of technologies to help motorists find unoccupied parking spaces, car location when returning to the vehicle and improve their experience. This includes adaptive lighting sensors and parking space led indicators (red for occupied, green for available and blue is reserved for the disabled; above every parking space), and indoor positioning system (IPS).

PGI systems are a product of the worldwide initiative for the development of intelligent transportation system in urban areas. PGI systems can assist in the development of safe, efficient and environmentally friendly transportation network.

PGI systems are designed to aid in the search for vacant parking spaces by directing drivers to car parks where occupancy levels are low. The objective is to reduce search time, which in turn reduces congestion on the surrounding roads for other traffic with related benefits to air pollution with the ultimate aim of enhancement of the urban area.

Parking Guidance System (PGS)

Ultrasonic sensors above each lot in this indoor car park determine if a car has already taken the lot and indicate using leds; and some can send a Bluetooth/SMS message with the parking space number or code.

A Parking Guidance System (PGS) have different elements:

- Detectors
- LED indicators
- Zone Controllers
- Data / Intermediate Controllers
- Central Control System
- Signs

Detectors

Operation of vehicle detection sensor and available space indicator. The indicator turns from green to red when the space becomes unavailable.

Detectors can use different technologies, such as:

- Ultrasonic detectors
- Forward mounting sensors.
- Camera based detectors.
- Infrared based detectors.
- Magnetic sensors that detect vehicles via changes in earth's magnetic field.
- Some use combinations of above technologies to detect vehicles.

LED Indicators

LED indicators. Some manufacturers use LED indicators per space, while others either don't provide any or use one LED indicator for few spaces. The purpose of an LED indicator is to provide guidance to parkers up to the last empty space.

Zone Controllers

The Zone control unit (ZCU) (aka Node Controller), is the middle layer of parking ultrasonic detector, which is to manage a group of ultrasonic detectors, detects then loops the information of the detectors and send the relevant information to a higher element or directly to central controller. One zone / node controller can support between 16 and 100 detectors depending on the manufacturer of Parking Guidance System.

Data / Intermediate Controllers

Some manufacturers use a kind of intermediate controllers, to accumulate data and then channel it to the Central Control Unit and others connect their Zone Controllers to connect directly to the Central Control Unit.

Central Control System

Parking Guidance System management software, an electronic map of the parking space is embedded in the software, which can directly reflect the usage situation of the carpark in real time. The operators can monitor the carpark situation according to the electronic map. As for the wrongly parked vehicles, the system supports manual rewriting its parking spaces to adjust the real occupied situation of the parking space. Database supporting is usually SQL Server/Access.

Signs

- Indoor and outdoor LED displays are used to inform drivers of the available parking spaces and guide them to empty ones.

Road Weather Information System

A Road Weather Information System (RWIS) comprises automatic weather stations (technically referred to as Environmental Sensor Stations (ESS)) in the field, a communication system for data transfer, and central systems to collect field data from numerous ESS. These stations measure real-time atmospheric, pavement, water level conditions, and visibility. Central RWIS hardware and software are used to process observations from ESS to develop nowcasts or forecasts, and display or disseminate road weather information in a format that can be easily interpreted by a manager. RWIS data are used by road operators and maintainers to support decision making. Real-time RWIS data is also used by Automated Warning Systems (AWS).

References

- David Lowe (2005). The Transport Manager's and Operator's Handbook 2006. Kogan Page Publishers. p. 239. ISBN 0-7494-4488-6.
- Hammoudi, K.; Benhabiles, H.; Jandial, A.; Dornaika, F.; Mouzna, J. (1 August 2016). "Developing a Vision-based Adaptive Parking Space Management System". International Journal of Sensors, Wireless Communications and Control. 6: 101–113. Retrieved 5 August 2016.
- "National ANPR Standards for Policing Part 3 – Data Access and Management Standards" (PDF). Home Office. October 2014. Retrieved 31 January 2016.
- "SchNEWS 625 - Sussex Police try to close down Smash EDO film, Big Brother Britain, Depleted Uranium raid, and more...". Schnews.org.uk. Retrieved 2016-06-30.
- "Speed-Measuring Device Performance Specifications:Across-the-Road Radar Module" (PDF). NHTSA.gov. Retrieved 2016-06-30.
- Idris, M.Y.I.; Leng, Y.Y.; Tamil, E.M.; Noor, N.M.; Razak, Z. (1 February 2009). "Car Park System: A Review of Smart Parking System and its Technology". Information Technology Journal. 8 (2): 101–113. doi:10.3923/itj.2009.101.113. Retrieved 24 September 2015.
- "1993 Eunos/Mazda Cosmo Classic Drive Uncosmopolitan: Meet the Rarest Mazda in America". Motor Trend. TEN: The Enthusiast Network. Retrieved 2015-01-18.
- "Shan Du ; IntelliView Technol., Inc., Calgary, AB, Canada ; Ibrahim, M. ; Shehata, M. ; Badawy, Wael; Automatic License Plate Recognition (ALPR): A State-of-the-Art Review". IEEE. 1 Feb 2013. Retrieved 2014-01-09.
- Kwaśnicka, Halina; Wawrzyniak, Bartosz (2002), "License plate localization and recognition in camera pictures" (PDF), AI-METH 2002 – Artificial Intelligence Methods November 13–15, Gliwice, Poland, retrieved 2014-01-13
- the license plate recognition company (2004-10-20). "License Plate Recognition Algorithms and Technology". Platerecognition.info. Retrieved 2014-04-23.
- Hansen, Megan and Will Jason (27 Mar 2013). "Golden Gate Bridge sees smooth transition to all-electronic tolls". Marin Independent Journal. Retrieved 17 May 2013.
- Friedrich, Markus; Jehlicka, Prokop; Schlaich, Johannes (2008). "Automatic number plate recognition for the observance of travel behavior". Universität Stuttgart Institut für Straßen und Verkehrswesen. Retrieved 2013-07-02.

5

Transportation Forecasting Model and its Components

Transportation planning and engineering requires real-time movement of traffic and the use of transportation facilities for future road as well as urban transport and manufacture. It also prominently features in environmental conservation and urban pollution. The aspects elucidated in this chapter are of vital importance, and provide a better understanding of transportation and vehicular engineering.

Transportation Forecasting

Transportation forecasting is the attempt of estimating the number of vehicles or people that will use a specific transportation facility in the future. For instance, a forecast may estimate the number of vehicles on a planned road or bridge, the ridership on a railway line, the number of passengers visiting an airport, or the number of ships calling on a seaport. Traffic forecasting begins with the collection of data on current traffic. This traffic data is combined with other known data, such as population, employment, trip rates, travel costs, etc., to develop a traffic demand model for the current situation. Feeding it with predicted data for population, employment, etc. results in estimates of future traffic, typically estimated for each segment of the transportation infrastructure in question, e.g., for each roadway segment or railway station.

Traffic forecasts are used for several key purposes in transportation policy, planning, and engineering: to calculate the capacity of infrastructure, e.g., how many lanes a bridge should have; to estimate the financial and social viability of projects, e.g., using cost–benefit analysis and social impact assessment; and to calculate environmental impacts, e.g., air pollution and noise.

Four-step Models

Within the rational planning framework, transportation forecasts have traditionally followed the sequential four-step model or urban transportation planning (UTP) procedure, first implemented on mainframe computers in the 1950s at the Detroit Metropolitan Area Traffic Study and Chicago Area Transportation Study (CATS).

Land-use forecasting starts the process. Typically, forecasts are made for the region as a whole, e.g., of population growth. Such forecasts provide control totals for the local land use analysis. Typically, the region is divided into zones and by trend or regression analysis, the population and employment are determined for each.

The four steps of the classical urban transportation planning system model are:

- Trip generation determines the frequency of origins or destinations of trips in each zone by trip purpose, as a function of land uses and household demographics, and other socio-economic factors.

- Trip distribution matches origins with destinations, often using a gravity model function, equivalent to an entropy maximizing model. Older models include the fratar model.
- Mode choice computes the proportion of trips between each origin and destination that use a particular transportation mode. (This modal model may be of the logit form, developed by Nobel Prize winner Daniel McFadden.)
- Route assignment allocates trips between an origin and destination by a particular mode to a route. Often (for highway route assignment) Wardrop's principle of user equilibrium is applied (equivalent to a Nash equilibrium), wherein each driver (or group) chooses the shortest (travel time) path, subject to every other driver doing the same. The difficulty is that travel times are a function of demand, while demand is a function of travel time, the so-called bi-level problem. Another approach is to use the Stackelberg competition model, where users ("followers") respond to the actions of a "leader", in this case for example a traffic manager. This leader anticipates on the response of the followers.

After the classical model, there is an evaluation according to an agreed set of decision criteria and parameters. A typical criterion is cost–benefit analysis. Such analysis might be applied after the network assignment model identifies needed capacity: is such capacity worthwhile? In addition to identifying the forecasting and decision steps as additional steps in the process, it is important to note that forecasting and decision-making permeate each step in the UTP process. Planning deals with the future, and it is forecasting dependent.

Activity-based Models

Activity-based models are another class of models that predict for individuals where and when specific activities (e.g. work, leisure, shopping, ...) are conducted.

Partial and fully operational activity-based simulation systems include:

- STARCHILD
- MIDAS (Micro-analytic Integrated Demographic Accounting System)
- CT-RAMP
- CEMDAP
- Daysim
- PCATS (Prism Constrained Activity-Travel Simulator)
- SIMAP
- Polaris Github
- ALBATROSS ("A Learning-Based Transportation Oriented Simulation System") model.
- FAMOS (Florida's Activity Mobility Simulator)

- TASHA (Travel Activity Scheduler for Household agents)
- TAPAS (Travel and Activity Patterns Simulation)
- the Best Practice Models of the New York Metropolitan Transportation Council, Columbus and San Francisco County
- MATSIM-T (Multi-Agent Transport Simulation Toolkit).
- FEATHERS (Forecasting Evolutionary Activity-Travel of Households and their Environmental RepercussionS).

The major premise behind activity-based models is that travel demand is derived from activities that people need or wish to perform, with travel decisions forming part of the scheduling decisions. Travel is then seen as just one of the attributes of a system. The travel model is therefore set within the context of an agenda, as a component of an activity scheduling decision.

Activity-based models offer other possibilities than four-step models, e.g. to model environmental issues such as emissions and exposure to air pollution. Although their obvious advantages for environmental purposes were recognized by Shiftan almost a decade ago, applications to exposure models remain scarce. Activity-based models have recently been used to predict emissions and air quality. They can also provide a better total estimate of exposure while also enabling the disaggregation of individual exposure over activities. They can therefore be used to reduce exposure misclassification and establish relationships between health impacts and air quality more precisely. Policy makers can use activity-based models to devise strategies that reduce exposure by changing time activity patterns or that target specific groups in the population.

Precursor Steps

Although not identified as steps in the UTP process, a lot of data gathering is involved in the UTP analysis process. Census and land use data are obtained, along with home interview surveys and journey surveys. Home interview surveys, land use data, and special trip attraction surveys provide the information on which the UTP analysis tools are exercised.

Data collection, management, and processing; model estimation; and use of models to yield plans are much used techniques in the UTP process. In the early days, in the USA, census data was augmented that with data collection methods that had been developed by the Bureau of Public Roads (a predecessor of the Federal Highway Administration): traffic counting procedures, cordon "where are you coming from and where are you going" counts, and home interview techniques. Protocols for coding networks and the notion of analysis or traffic zones emerged at the CATS.

Model estimation used existing techniques, and plans were developed using whatever models had been developed in a study. The main difference between now and then is the development of some analytic resources specific to transportation planning, in addition to the BPR data acquisition techniques used in the early days.

Critique

The sequential and aggregate nature of transportation forecasting has come under much criticism. While improvements have been made, in particular giving an activity-base to travel demand, much remains to be done. In the 1990s, most federal investment in model research went to the Transims project at Los Alamos National Laboratory, developed by physicists. While the use of supercomputers and the detailed simulations may be an improvement on practice, they have yet to be shown to be better (more accurate) than conventional models. A commercial version was spun off to IBM, and an open source version is also being actively maintained as TRANSIMS Open-Source.

One of the major oversights in the use of transportation models in practice is the absence of any feedback from transportation models on land use. Highways and transit investments not only respond to land use, they shape it as well.

Trip Generation

Trip generation is the first step in the conventional four-step transportation forecasting process (followed by trip distribution, mode choice, and route assignment), widely used for forecasting travel demands. It predicts the number of trips originating in or destined for a particular traffic analysis zone.Guidelines for trip generation analysis

Typically, trip generation analysis focuses on residences, and residential trip generation is thought of as a function of the social and economic attributes of households. At the level of the traffic analysis zone, residential land uses “produce” or generate trips. Traffic analysis zones are also destinations of trips, trip attractors. The analysis of attractors focuses on nonresidential land uses.

Input Data

A forecasting activity, such as one based on the concept of economic base analysis, provides aggregate measures of population and activity growth. Land use forecasting distributes forecast changes in activities in a disaggregate-spatial manner among zones. The next step in the transportation planning process addresses the question of the frequency of origins and destinations of trips in each zone: for short, trip generation.

Early Analysis

The first zonal trip generation (and its inverse, attraction) analysis in the Chicago Area Transportation Study (CATS) followed the “decay of activity intensity with distance from the central business district (CBD)” thinking current at the time. Data from extensive surveys were arrayed and interpreted on a-distance-from-CBD scale. For example, a commercial land use in ring 0 (the CBD and vicinity) was found to generate 728 vehicle trips per day in 1956. That same land use in ring 5 (about 17 km (11 mi) from the CBD) generated about 150 trips per day.

The case of trip destinations will illustrate use of the concept of activity decline with intensity (as measured by distance from CBD) worked. Destination data are arrayed:

Table: Trip Destinations per unit (Acre) of Land				
Ring	**Manufacturing**	**Commercial**	**Open Space**	**etc.**
0	X_{1m}	X_{1c}	etc.	
.	.	.	.	
.	.	.	.	
7	x_{7m}	x_{7c}	etc.	

The land use analysis provides information on how land uses will change from an initial year (say t = 0) to some forecast year (say t = 20). Suppose we are examining a zone. We take the mix of land uses projected, say, for year t = 20 and apply the trip destination rates for the ring in which the zone is located. That is, there will this many acres of commercial land use, that many acres of public open space, etc., in the zone. The acres of each use type are multiplied by the ring specific destination rates. The result is summed to yield the zone's trip destinations. It is to be noted that the CATS assumed that trip destination rates would not change over time.

Later Analysis

As was true for land use analysis, the approach developed at CATS was considerably modified in later studies. The conventional four-step paradigm evolved as follows: Types of trips are considered. Home-based (residential) trips are divided into work and other, with major attention given to work trips. Movement associated with the home end of a trip is called trip production, whether the trip is leaving or coming to the home. Non-home-based or non-residential trips are those a home base is not involved. In this case, the term production is given to the origin of a trip and the term attraction refers to the destination of the trip.

Residential trip generation analysis is often undertaken using statistical regression. Person, transit, walking, and auto trips per unit of time are regressed on variables thought to be explanatory, such as: household size, number of workers in the household, persons in an age group, type of residence (single family, apartment, etc.), and so on. Usually, measures on five to seven independent variables are available; additive causality is assumed.

Usually also, regressions are made at the aggregate/zone level. Variability among households within a zone isn't measured when data are aggregated. High correlation coefficients are found when regressions are run on aggregate data, say, about 0.90, but lower coefficients, say, about 0.25, are found when regressions are made on observation units such as households. In short, there is much variability that is hidden by aggregation.

Sometimes cross-classification techniques are applied to residential trip generation problems. The CATS procedure described above is a cross-classification procedure.

Classification techniques are often used for non-residential trip generation. First, the type of land use is a factor influencing travel, it is regarded as a causal factor. A list of land uses and associated trip rates illustrated a simple version of the use of this technique:

Table: Trips per day	
Land Use Type	**Trips**
Department Store	X
Grocery Store	Y
etc.	

Such a list can be improved by adding information. Large, medium, and small might be defined for each activity and rates given by size. Number of employees might be used: for example, <10, 10-20, etc. Also, floor space is used to refine estimates.

In other cases, regressions, usually of the form trip rate = f(number of employees, floor area of establishment), are made for land use types.

Special treatment is often given major trip generators: large shopping centers, airports, large manufacturing plants, and recreation facilities.

The theoretical work related to trip generation analysis is grouped under the rubric travel demand theory, which treats trip generation-attraction, as well as mode choice, route selection, and other topics.

ITE Trip Generation Procedures

The Institute of Transportation Engineers's Trip Generation informational report provides trip generation rates for numerous land use and building types. The planner can add local adjustment factors and treat mixes of uses with ease. Ongoing work is adding to the stockpile of numbers; over 4000 studies were aggregated for the current edition.

ITE Procedures estimate the number of trips entering or exiting a site at a given time (sometimes the number entering and exiting combined is estimated). ITE Rates are functions of type of development, and square footage, number of gas pumps, number of dwelling units, or other standard measurable things, usually produced in site plans. They are typically of the form $Trips = a + b * Area$ OR $Trips = a + bln(Area)$. They do not consider location, competitors, complements, the cost of transportation, or many other obviously likely important factors. They are often estimated based on very few observations (a non-statistically significant sample). Many localities require their use to ensure adequate public facilities for growth management and subdivision approval.

Trip Distribution

Trip distribution (or destination choice or zonal interchange analysis) is the second component (after trip generation, but before mode choice and route assignment) in the traditional four-step transportation forecasting model. This step matches tripmakers' origins and destinations to develop a "trip table", a matrix that displays the number of trips going from each origin to each destination. Historically, this component has been the least developed component of the transportation planning model.

Table: Illustrative trip table				
Origin \ Destination	**1**	**2**	**3**	**Z**
1	T_{11}	T_{12}	T_{13}	T_{1Z}
2	T_{21}			
3	T_{31}			
Z	T_{Z1}			T_{ZZ}

Where: T_{ij} = trips from origin *i* to destination *j*. Note that the practical value of trips on the diagonal, e.g. from zone 1 to zone 1, is zero since no intra-zonal trip occurs.

Work trip distribution is the way that travel demand models understand how people take jobs. There are trip distribution models for other (non-work) activities such as the choice of location for grocery shopping, which follow the same structure.

History

Over the years, modelers have used several different formulations of trip distribution. The first was the Fratar or Growth model (which did not differentiate trips by purpose). This structure extrapolated a base year trip table to the future based on growth, but took no account of changing spatial accessibility due to increased supply or changes in travel patterns and congestion. (Simple Growth factor model, Furness Model and Detroit model are models developed at the same time period)

The next models developed were the gravity model and the intervening opportunities model. The most widely used formulation is still the gravity model.

While studying traffic in Baltimore, Maryland, Alan Voorhees developed a mathematical formula to predict traffic patterns based on land use. This formula has been instrumental in the design of numerous transportation and public works projects around the world. He wrote "A General Theory of Traffic Movement," (Voorhees, 1956) which applied the gravity model to trip distribution, which translates trips generated in an area to a matrix that identifies the number of trips from each origin to each destination, which can then be loaded onto the network.

Evaluation of several model forms in the 1960s concluded that "the gravity model and intervening opportunity model proved of about equal reliability and utility in simulating the 1948 and 1955 trip distribution for Washington, D.C." (Heanue and Pyers 1966). The Fratar model was shown to have weakness in areas experiencing land use changes. As comparisons between the models showed that either could be calibrated equally well to match observed conditions, because of computational ease, gravity models became more widely spread than intervening opportunities models. Some theoretical problems with the intervening opportunities model were discussed by Whitaker and West (1968) concerning its inability to account for all trips generated in a zone which makes it more difficult to calibrate, although techniques for dealing with the limitations have been developed by Ruiter (1967).

With the development of logit and other discrete choice techniques, new, demographically disaggregate approaches to travel demand were attempted. By including variables other than travel time in determining the probability of making a trip, it is expected to have a better prediction of travel behavior. The logit model and gravity model have been shown by Wilson (1967) to be of essentially the same form as used in statistical mechanics, the entropy maximization model. The application of these models differs in concept in that the gravity model uses impedance by travel time, perhaps stratified by socioeconomic variables, in determining the probability of trip making, while a discrete choice approach brings those variables inside the utility or impedance function. Discrete choice models require more information to estimate and more computational time.

Ben-Akiva and Lerman (1985) have developed combination destination choice and mode choice models using a logit formulation for work and non-work trips. Because of computational intensity, these formulations tended to aggregate traffic zones into larger districts or rings in estimation. In current application, some models, including for instance the transportation planning model used in Portland, Oregon, use a logit formulation for destination choice. Allen (1984) used utilities from a logit based mode choice model in determining composite impedance for trip distribution. However, that approach, using mode choice log-sums implies that destination choice depends on the same variables as mode choice. Levinson and Kumar (1995) employ mode choice probabilities as a weighting factor and develop a specific impedance function or "f-curve" for each mode for work and non-work trip purposes.

Mathematics

At this point in the transportation planning process, the information for zonal interchange analysis is organized in an origin-destination table. On the left is listed trips produced in each zone. Along the top are listed the zones, and for each zone we list its attraction. The table is n x n, where n = the number of zones.

Each cell in our table is to contain the number of trips from zone i to zone j. We do not have these within-cell numbers yet, although we have the row and column totals. With data organized this way, our task is to fill in the cells for tables headed $t = 1$ through say $t = n$.

Actually, from home interview travel survey data and attraction analysis we have the cell information for $t = 1$. The data are a sample, so we generalize the sample to the universe. The techniques used for zonal interchange analysis explore the empirical rule that fits the $t = 1$ data. That rule is then used to generate cell data for $t = 2$, $t = 3$, $t = 4$, etc., to $t = n$.

The first technique developed to model zonal interchange involves a model such as this:

$$T_{ij} = T_i \frac{A_j f\left(C_{ij}\right) K_{ij}}{\sum_{j=1}^{n} A_j f\left(C_{ij}\right) K_{ij}}$$

where:

- T_{ij} : trips from i to j.
- T_i : trips from i, as per our generation analysis
- A_j : trips attracted to j, as per our generation analysis

- $f\left(C_{ij}\right)$: travel cost friction factor, say $= C_{ij}^{b}$
- A_j : Calibration parameter

Zone *i* generates T_i trips; how many will go to zone *j*? That depends on the attractiveness of *j* compared to the attractiveness of all places; attractiveness is tempered by the distance a zone is from zone *i*. We compute the fraction comparing *j* to all places and multiply $T_{;i}$ by it.

The rule is often of a gravity form:

$$T_{ij} = a\frac{P_i P_j}{C_{ij}^{b}}$$

where:

- $P_i P_j$: populations of *i* and *j*
- $a;b$: parameters

But in the zonal interchange mode, we use numbers related to trip origins ($T_{;i}$) and trip destinations ($T_{;j}$) rather than populations.

There are lots of model forms because we may use weights and special calibration parameters, e.g., one could write say:

$$T_{ij} = a\frac{T_i^{c} T_j^{d}}{C_{ij}^{b}}$$

Or

$$T_{ij} = \frac{cT_i dT_j}{C_{ij}^{b}}$$

where:

- *a, b, c, d* are parameters
- C_{ij} : travel cost (e.g. distance, money, time)
- T_j : inbound trips, destinations
- T_i : outbound trips, origin

Gravity Model

The gravity model illustrates the macroscopic relationships between places (say homes and workplaces). It has long been posited that the interaction between two locations declines with increasing (distance, time, and cost) between them, but is positively associated with the amount of activity at each location (Isard, 1956). In analogy with physics, Reilly (1929) formulated Reilly's law of retail gravitation, and J. Q. Stewart (1948) formulated definitions of demographic gravitation, force, energy, and potential, now called accessibility (Hansen, 1959). The distance decay factor of 1/distance has been updated to a more comprehensive function of generalized cost, which is not necessarily linear - a negative exponential tends to be the preferred form.

The gravity model has been corroborated many times as a basic underlying aggregate relationship (Scott 1988, Cervero 1989, Levinson and Kumar 1995). The rate of decline of the interaction (called alternatively, the impedance or friction factor, or the utility or propensity function) has to be empirically measured, and varies by context.

Limiting the usefulness of the gravity model is its aggregate nature. Though policy also operates at an aggregate level, more accurate analyses will retain the most detailed level of information as long as possible. While the gravity model is very successful in explaining the choice of a large number of individuals, the choice of any given individual varies greatly from the predicted value. As applied in an urban travel demand context, the disutilities are primarily time, distance, and cost, although discrete choice models with the application of more expansive utility expressions are sometimes used, as is stratification by income or vehicle ownership.

Mathematically, the gravity model often takes the form:

$$T_{ij} = K_i K_j T_i T_j f(C_{ij})$$

$$\sum_j T_{ij} = T_i, \sum_i T_{ij} = T_j$$

$$K_i = \frac{1}{\sum_j K_j T_j f(C_{ij})}, K_j = \frac{1}{\sum_i K_i T_i f(C_{ij})}$$

where

- T_{ij} = Trips between origin i and destination j
- T_i = Trips originating at i
- T_j = Trips destined for j
- C_{ij} = travel cost between i and j
- K_i, K_j = balancing factors solved iteratively.
- f = distance decay factor, as in the accessibility model

It is doubly constrained, in the sense that for any i the total number of trips from i predicted by the model always (mechanically, for any parameter values) equals the real total number of trips from i. Similarly, the total number of trips to j predicted by the model equals the real total number of trips to j, for any j.

Entropy Analysis

Wilson (1970) gives another way to think about zonal interchange problem. This section treats Wilson's methodology to give a grasp of central ideas.

To start, consider some trips where there are seven people in origin zones commuting to seven jobs in destination zones. One configuration of such trips will be:

Table: Configuration of trips			
zone	1	2	3
1	2	1	1
2	0	2	1

$$w\left(T_{ij}\right)=\frac{7!}{2!1!1!0!2!1!}=1260$$

where 0! = 1.

That configuration can appear in 1,260 ways. We have calculated the number of ways that configuration of trips might have occurred, and to explain the calculation, let's recall those coin tossing experiments talked about so much in elementary statistics.

The number of ways a two-sided coin can come up is 2^n, where n is the number of times we toss the coin. If we toss the coin once, it can come up heads or tails, $2^1=2$. If we toss it twice, it can come up HH, HT, TH, or TT, four ways, and $2^2=4$. To ask the specific question about, say, four coins coming up all heads, we calculate $4!/(4!0!)=1$. Two heads and two tails would be $4!/(2!2!)=6$. . We are solving the equation:

$$w=\frac{n!}{\prod_{i=1}^{n} n_i!}$$

An important point is that as *n* gets larger, our distribution gets more and more peaked, and it is more and more reasonable to think of a most likely state.

However, the notion of most likely state comes not from this thinking; it comes from statistical mechanics, a field well known to Wilson and not so well known to transportation planners. The result from statistical mechanics is that a descending series is most likely. Think about the way the energy from lights in the classroom is affecting the air in the classroom. If the effect resulted in an ascending series, many of the atoms and molecules would be affected a lot and a few would be affected a little. The descending series would have many affected not at all or not much and only a few affected very much. We could take a given level of energy and compute excitation levels in ascending and descending series. Using the formula above, we would compute the ways particular series could occur, and we would conclude that descending series dominate.

That is more-or-less Boltzmann's Law,

$$p_j=p_0e^{\beta e_j}$$

That is, the particles at any particular excitation level *j* will be a negative exponential function of the particles in the ground state, p_0, the excitation level, e_j, and a parameter β, which is a function of the (average) energy available to the particles in the system.

The two paragraphs above have to do with ensemble methods of calculation developed by Gibbs, a topic well beyond the reach of these notes.

Returning to the O-D matrix, note that we have not used as much information as we would have from an O and D survey and from our earlier work on trip generation. For the same travel pattern in the O-D matrix used before, we would have row and column totals, i.e.:

Table: Illustrative O-D matrix with row and column totals

	zone	1	2	3
zone	$T_i \backslash T_j$	2	3	2
1	4	2	1	1
2	3	0	2	1

Consider the way the four folks might travel, 4!/(2!1!1!) = 12; consider three folks, 3!/(0!2!1!) = 3. All travel can be combined in 12×3 = 36 ways. The possible configuration of trips is, thus, seen to be much constrained by the column and row totals.

We put this point together with the earlier work with our matrix and the notion of most likely state to say that we want to

$$\max w\left(T_{ij}\right) = \frac{T!}{\prod_{ij} Tij!}$$

subject to

$$\sum_j T_{ij} = T_i; \sum_i T_{ij} = T_j$$

where:

$$T = \sum_j \sum_i T_{ij} = \sum_i T_i = \sum_j T_j$$

and this is the problem that we have solved above.

Wilson adds another consideration; he constrains the system to the amount of energy available (i.e., money), and we have the additional constraint,

$$\sum_i \sum_j T_{ij} C_{ij} = C$$

where *C* is the quantity of resources available and C_{ij} is the travel cost from *i* to *j*.

The discussion thus far contains the central ideas in Wilson's work, but we are not yet to the place where the reader will recognize the model as it is formulated by Wilson.

First, writing the Λ function to be maximized using Lagrangian multipliers, we have:

$$(T_{ij},\lambda_i,\lambda_j)=\frac{T!}{\prod_{ij}Tij!}+\sum_i\lambda_i\left(T_i-\sum_j T_{ij}\right)+\sum_j\lambda_j\left(T_j-\sum_i T_{ij}\right)+\beta\left(C-\sum_i\sum_j T_{ij}C_{ij}\right)$$

where λ_i, λ_j and β are the Lagrange multipliers, β having an energy sense.

Second, it is convenient to maximize the natural log (ln) rather than $w(T_{ij})$, for then we may use Stirling's approximation.

$$\ln N! \approx N \ln N - N$$

So,

$$\frac{\partial \ln N!}{\partial N} \approx \ln N$$

Third, evaluating the maximum, we have

$$\frac{\partial\Lambda(T_{ij},\lambda_i,\lambda_j)}{\partial T_{ij}} = -\ln T_{ij}-\lambda_i-\lambda_j-\beta C_{ij}=0$$

with solution

$$\ln T_{ij} = -\lambda_i-\lambda_j-\beta C_{ij}$$

$$T_{ij} = e^{-\lambda_i-\lambda_j-\beta C_{ij}}$$

Finally, substituting this value of T_{ij} back into our constraint equations, we have:

$$\sum_j e^{-\lambda_i-\lambda_j-\beta C_{ij}} = T_i;\sum_i e^{-\lambda_i-\lambda_j-\beta C_{ij}} = T_j$$

and, taking the constant multiples outside of the summation sign

$$e^{-\lambda_i} = \frac{T_i}{\sum_j e^{-\lambda_j-\beta C_{ij}}};e^{-\lambda_j} = \frac{T_j}{\sum_i e^{-\lambda_i-\beta C_{ij}}}$$

Let

$$\frac{e^{-\lambda_i}}{T_i} = A_i;\frac{e^{-\lambda_j}}{T_j} = B_j$$

we have

$$T_{ij} = A_i B_j T_i T_j e^{-\beta C_{ij}}$$

which says that the most probable distribution of trips has a gravity model form, T_{ij} is proportional to trip origins and destinations. The constants A_i , B_j , and β ensure that constraints are met.

Turning now to computation, we have a large problem. First, we do not know the value of C, which earlier on we said had to do with the money available, it was a cost constraint. Consequently, we have to set β to different values and then find the best set of values for A_i and B_j . We know what β means – the greater the value of β , the less the cost of average distance traveled. (Compare β in Boltzmann's Law noted earlier.) Second, the values of A_i and B_j Issues

Congestion

One of the key drawbacks to the application of many early models was the inability to take account of congested travel time on the road network in determining the probability of making a trip between two locations. Although Wohl noted as early as 1963 research into the feedback mechanism or the "interdependencies among assigned or distributed volume, travel time (or travel 'resistance') and route or system capacity", this work has yet to be widely adopted with rigorous tests of convergence, or with a so-called "equilibrium" or "combined" solution (Boyce et al. 1994). Haney (1972) suggests internal assumptions about travel time used to develop demand should be consistent with the output travel times of the route assignment of that demand. While small methodological inconsistencies are necessarily a problem for estimating base year conditions, forecasting becomes even more tenuous without an understanding of the feedback between supply and demand. Initially heuristic methods were developed by Irwin and Von Cube (as quoted in Florian et al. (1975)) and others, and later formal mathematical programming techniques were established by Evans (1976).

Stability of Travel Times

A key point in analyzing feedback is the finding in earlier research by Levinson and Kumar (1994) that commuting times have remained stable over the past thirty years in the Washington Metropolitan Region, despite significant changes in household income, land use pattern, family structure, and labor force participation. Similar results have been found in the Twin Cities by Barnes and Davis (2000).

The stability of travel times and distribution curves over the past three decades gives a good basis for the application of aggregate trip distribution models for relatively long term forecasting. This is not to suggest that there exists a constant travel time budget.

Mode Choice

Mode choice analysis is the third step in the conventional four-step transportation forecasting model. The steps, in order, are trip generation, trip distribution, mode choice analysis, and route assignment. Trip distribution's zonal interchange analysis yields a set of origin destination tables that tells where the trips will be made. Mode choice analysis allows the modeler to determine what mode of transport will be used, and what modal share results.

The early transportation planning model developed by the Chicago Area Transportation Study (CATS) focused on transit. It wanted to know how much travel would continue by transit. The CATS divided transit trips into two classes: trips to the Central Business District, or CBD (mainly by subway/elevated transit, express buses, and commuter trains) and other (mainly on the local bus system). For the latter, increases in auto ownership and use were a trade-off against bus use; trend data were used. CBD travel was analyzed using historic mode choice data together with projections of CBD land uses. Somewhat similar techniques were used in many studies. Two decades after CATS, for example, the London study followed essentially the same procedure, but in this case, researchers first divided trips into those made in the inner part of the city and those in the outer part. This procedure was followed because it was thought that income (resulting in the purchase and use of automobiles) drove mode choice.

Diversion Curve Techniques

The CATS had diversion curve techniques available and used them for some tasks. At first, the CATS studied the diversion of auto traffic from streets and arterial roads to proposed expressways. Diversion curves were also used for bypasses built around cities to find out what percent of traffic would use the bypass. The mode choice version of diversion curve analysis proceeds this way: one forms a ratio, say:

$$\frac{c_{\text{transit}}}{c_{\text{auto}}} = R$$

where:

c_m = travel time by mode m and

R is empirical data in the form:

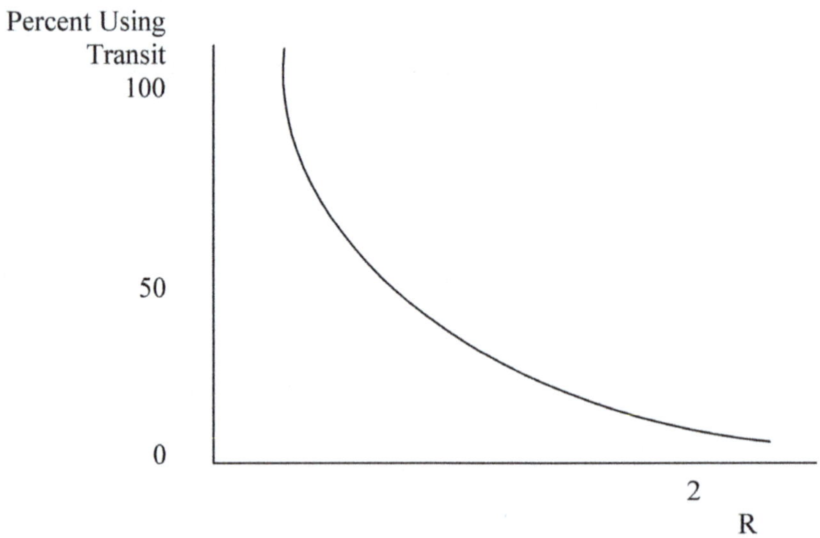

Given the *R* that we have calculated, the graph tells us the percent of users in the market that will choose transit. A variation on the technique is to use costs rather than time in the diversion ratio. The decision to use a time or cost ratio turns on the problem at hand. Transit agencies developed diversion curves for different kinds of situations, so variables like income and population density entered implicitly.

Diversion curves are based on empirical observations, and their improvement has resulted from better (more and more pointed) data. Curves are available for many markets. It is not difficult to obtain data and array results. Expansion of transit has motivated data development by operators and planners. Yacov Zahavi's UMOT studies, discussed earlier, contain many examples of diversion curves.

In a sense, diversion curve analysis is expert system analysis. Planners could "eyeball" neighborhoods and estimate transit ridership by routes and time of day. Instead, diversion is observed empirically and charts drawn.

Disaggregate Travel Demand Models

Travel demand theory was introduced in the appendix on traffic generation. The core of the field is the set of models developed following work by Stan Warner in 1962 (Strategic Choice of Mode in Urban Travel: A Study of Binary Choice). Using data from the CATS, Warner investigated classification techniques using models from biology and psychology. Building from Warner and other early investigators, disaggregate demand models emerged. Analysis is disaggregate in that individuals are the basic units of observation, yet aggregate because models yield a single set of parameters describing the choice behavior of the population. Behavior enters because the theory made use of consumer behavior concepts from economics and parts of choice behavior concepts from psychology. Researchers at the University of California, Berkeley (especially Daniel McFadden, who won a Nobel Prize in Economics for his efforts) and the Massachusetts Institute of Technology (Moshe Ben-Akiva) (and in MIT associated consulting firms, especially Cambridge Systematics) developed what has become known as choice models, direct demand models (DDM), Random Utility Models (RUM) or, in its most used form, the multinomial logit model (MNL).

Choice models have attracted a lot of attention and work; the Proceedings of the International Association for Travel Behavior Research chronicles the evolution of the models. The models are treated in modern transportation planning and transportation engineering textbooks.

One reason for rapid model development was a felt need. Systems were being proposed (especially transit systems) where no empirical experience of the type used in diversion curves was available. Choice models permit comparison of more than two alternatives and the importance of attributes of alternatives. There was the general desire for an analysis technique that depended less on aggregate analysis and with a greater behavioral content. And there was attraction, too, because choice models have logical and behavioral roots extended back to the 1920s as well as roots in Kelvin Lancaster's consumer behavior theory, in utility theory, and in modern statistical methods.

Psychological Roots

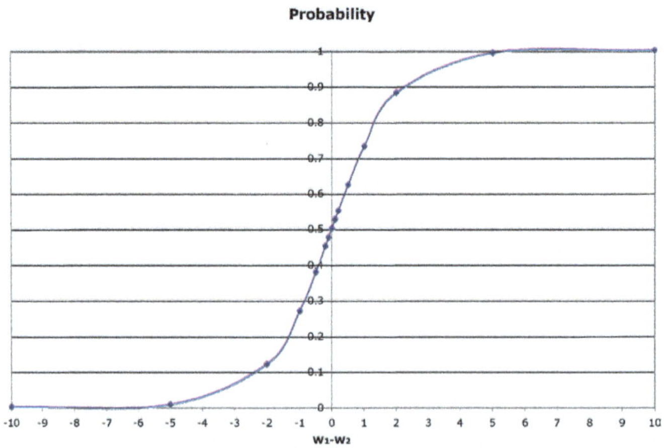

Early psychology work involved the typical experiment: Here are two objects with weights, w_1 and w_2, which is heavier? The finding from such an experiment would be that the greater the difference in weight, the greater the probability of choosing correctly. Graphs similar to the one on the right result.

Louis Leon Thurstone proposed (in the 1920s) that perceived weight,

$$w = v + e,$$

where v is the true weight and e is random with

$$E(e) = 0.$$

The assumption that e is normally and identically distributed (NID) yields the binary probit model.

Econometric formulation

Economists deal with utility rather than physical weights, and say that

observed utility = mean utility + random term.

The characteristics of the object, x, must be considered, so we have

$$u(x) = v(x) + e(x).$$

If we follow Thurston's assumption, we again have a probit model.

An alternative is to assume that the error terms are independently and identically distributed with a Weibull, Gumbel Type I, or double exponential distribution. (They are much the same, and differ slightly in their tails (thicker) from the normal distribution). This yields the multinomial logit model (MNL). Daniel McFadden argued that the Weibull had desirable properties compared to other distributions that might be used. Among other things, the error terms are normally and identically distributed. The logit model is simply a log ratio of the probability of choosing a mode to the probability of not choosing a mode.

$$\log\left(\frac{P_i}{1-P_i}\right) = v(x_i)$$

Observe the mathematical similarity between the logit model and the S-curves we estimated earlier, although here share increases with utility rather than time. With a choice model we are explaining the share of travelers using a mode (or the probability that an individual traveler uses a mode multiplied by the number of travelers).

The comparison with S-curves is suggestive that modes (or technologies) get adopted as their utility increases, which happens over time for several reasons. First, because the utility itself is a function of network effects, the more users, the more valuable the service, higher the utility associated with joining the network. Second because utility increases as user costs drop, which happens when fixed costs can be spread over more users (another network effect). Third technological advances, which occur over time and as the number of users increases, drive down relative cost.

An illustration of a utility expression is given:

$$\log\left(\frac{P_A}{1-P_A}\right)=\beta_0+\beta_1\left(c_A-c_T\right)+\beta_2\left(t_A-t_T\right)+\beta_3 I+\beta_4 N=v_A$$

where

P_i = Probability of choosing mode i.

P_A = Probability of taking auto

c_A, c_T = cost of auto, transit

t_A, t_T = travel time of auto, transit

I = income

N = Number of travelers

With algebra, the model can be translated to its most widely used form

$$\frac{P_A}{1-P_A}=e^{v_A}$$

$$P_A=e^{v_A}-P_A e^{v_A}$$

$$P_A\left(1+e^{v_A}\right)=e^{v_A}$$

$$P_A=\frac{e^{v_A}}{1+e^{v_A}}$$

It is fair to make two conflicting statements about the estimation and use of this model:

1. it's a "house of cards", and
2. used by a technically competent and thoughtful analyst, it's useful.

The "house of cards" problem largely arises from the utility theory basis of the model specification.

Broadly, utility theory assumes that (1) users and suppliers have perfect information about the market; (2) they have deterministic functions (faced with the same options, they will always make the same choices); and (3) switching between alternatives is costless. These assumptions don't fit very well with what is known about behavior. Furthermore, the aggregation of utility across the population is impossible since there is no universal utility scale.

Suppose an option has a net utility u_{jk} (option k, person j). We can imagine that having a systematic part v_{jk} that is a function of the characteristics of an object and person j, plus a random part e_{jk}, which represents tastes, observational errors and a bunch of other things (it gets murky here). (An object such as a vehicle does not have utility, it is characteristics of a vehicle that have utility.) The introduction of e lets us do some aggregation. As noted above, we think of observable utility as being a function:

$$v_A = \beta_0 + \beta_1(c_A - c_T) + \beta_2(t_A - t_T) + \beta_3 I + \beta_4 N$$

where each variable represents a characteristic of the auto trip. The value β_o is termed an alternative specific constant. Most modelers say it represents characteristics left out of the equation (e.g., the political correctness of a mode, if I take transit I feel morally righteous, so β_o may be negative for the automobile), but it includes whatever is needed to make error terms NID.

Econometric Estimation

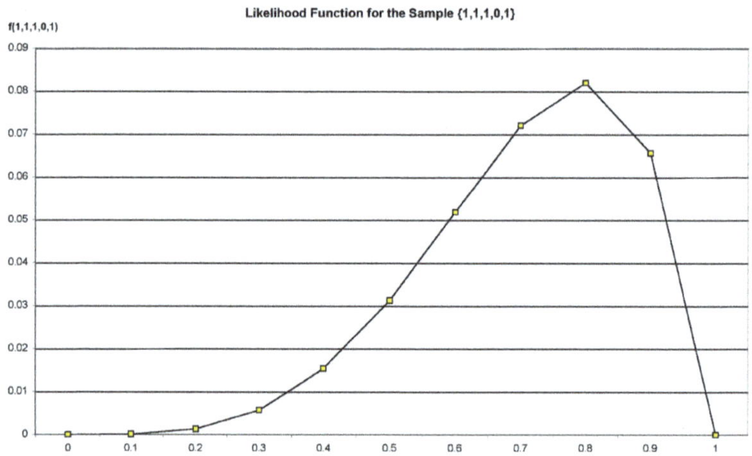

Turning now to some technical matters, how do we estimate $v(x)$? Utility ($v(x)$) isn't observable. All we can observe are choices (say, measured as 0 or 1), and we want to talk about probabilities of choices that range from 0 to 1. (If we do a regression on 0s and 1s we might measure for j a probability of 1.4 or −0.2 of taking an auto.) Further, the distribution of the error terms wouldn't have appropriate statistical characteristics.

The MNL approach is to make a maximum likelihood estimate of this functional form. The likelihood function is:

$$L^* = \prod_{n=1}^{N} f(y_n | x_n, \theta)$$

we solve for the estimated parameters

$$\hat{\theta}$$

that max L^*. This happens when:

$$\frac{\partial L}{\partial \hat{\theta}_N} = 0$$

The log-likelihood is easier to work with, as the products turn to sums:

$$\ln L^* = \sum_{n=1}^{N} \ln f\left(y_n \middle| x_n, \theta\right)$$

Consider an example adopted from John Bitzan's Transportation Economics Notes. Let X be a binary variable that is γ and 0 with probability (1 – *gamma*). Then f(0) = (1 – γ) and f(1) = γ. Suppose that we have 5 observations of X, giving the sample {1,1,1,0,1}. To find the maximum likelihood estimator of γ examine various values of γ, and for these values determine the probability of drawing the sample {1,1,1,0,1} If γ takes the value 0, the probability of drawing our sample is 0. If γ is 0.1, then the probability of getting our sample is: f(1,1,1,0,1) = f(1)f(1)f(1)f(0)f(1) = 0.1×0.1×0.1×0.9×0.1 = 0.00009 We can compute the probability of obtaining our sample over a range of γ – this is our likelihood function. The likelihood function for n independent observations in a logit model is

$$L^* = \prod_{n=1}^{N} P_i^{Y_i} \left(1 - P_i\right)^{1-Y_i}$$

where: Y_i = 1 or 0 (choosing e.g. auto or not-auto) and Pi = the probability of observing Y_i = 1

The log likelihood is thus:

$$\ell = \ln L^* = \sum_{i=1}^{n} \left[Y_i \ln P_i + (1 - Y_i) \ln (1 - P_i) \right]$$

In the binomial (two alternative) logit model,

$$P_{\text{auto}} = \frac{e^{v(x_{\text{auto}})}}{1 + e^{v(x_{\text{auto}})}}, so$$

$$\ell = \ln L^* = \sum_{i=1}^{n} \left[Y_i v(x_{\text{auto}}) - \ln\left(1 + e^{v(x_{\text{auto}})}\right) \right]$$

The log-likelihood function is maximized setting the partial derivatives to zero:

$$\frac{\partial \ell}{\partial \beta} = \sum_{i=1}^{n} \left(Y_i - \hat{P}_i \right) = 0$$

The above gives the essence of modern MNL choice modeling.

Additional Topics

Topics not touched on include the "red bus, blue bus" problem; the use of nested models (e.g., estimate choice between auto and transit, and then estimate choice between rail and bus transit); how consumers' surplus measurements may be obtained; and model estimation, goodness of fit, etc.

Returning to Roots

The discussion above is based on the economist's utility formulation. At the time MNL modeling was developed there was some attention to psychologist's choice work (e.g., Luce's choice axioms discussed in his Individual Choice Behavior, 1959). It has an analytic side in computational process modeling. Emphasis is on how people think when they make choices or solve problems. Put another way, in contrast to utility theory, it stresses not the choice but the way the choice was made. It provides a conceptual framework for travel choices and agendas of activities involving considerations of long and short term memory, effectors, and other aspects of thought and decision processes. It takes the form of rules dealing with the way information is searched and acted on. Although there is a lot of attention to behavioral analysis in transportation work, the best of modern psychological ideas are only beginning to enter the field. (e.g. Golledge, Kwan and Garling 1984; Garling, Kwan, and Golledge 1994).

Route Assignment

Route assignment, route choice, or traffic assignment concerns the selection of routes (alternative called paths) between origins and destinations in transportation networks. It is the fourth step in the conventional transportation forecasting model, following trip generation, trip distribution, and mode choice. The zonal interchange analysis of trip distribution provides origin-destination trip tables. Mode choice analysis tells which travelers will use which mode. To determine facility needs and costs and benefits, we need to know the number of travelers on each route and link of the network (a route is simply a chain of links between an origin and destination). We need to undertake traffic (or trip) assignment. Suppose there is a network of highways and transit systems and a proposed addition. We first want to know the present pattern of traffic delay and then what would happen if the addition were made.

Auto Assignment

Long-standing Techniques

The problem of estimating how many users are on each route is long standing. Planners started looking hard at it as freeways and expressways began to be developed. The freeway offered a superior level of service over the local street system, and diverted traffic from the local system. At first, diversion was the technique. Ratios of travel time were used, tempered by considerations of costs, comfort, and level of service.

The Chicago Area Transportation Study (CATS) researchers developed diversion curves for freeways versus local streets. There was much work in California also, for California had early experiences with freeway planning. In addition to work of a diversion sort, the CATS attacked some technical problems that arise when one works with complex networks. One result was the Bellman–Ford–Moore algorithm for finding shortest paths on networks.

The issue the diversion approach didn't handle was the feedback from the quantity of traffic on links and routes. If a lot of vehicles try to use a facility, the facility becomes congested and travel time increases. Absent some way to consider feedback, early planning studies (actually, most in the period 1960-1975) ignored feedback. They used the Moore algorithm to determine shortest paths and assigned all traffic to shortest paths. That's called all or nothing assignment because either all of the traffic from i to j moves along a route or it does not.

The all-or-nothing or shortest path assignment is not trivial from a technical-computational view. Each traffic zone is connected to $n - 1$ zones, so there are numerous paths to be considered. In addition, we are ultimately interested in traffic on links. A link may be a part of several paths, and traffic along paths has to be summed link by link.

An argument can be made favoring the all-or-nothing approach. It goes this way: The planning study is to support investments so that a good level of service is available on all links. Using the travel times associated with the planned level of service, calculations indicate how traffic will flow once improvements are in place. Knowing the quantities of traffic on links, the capacity to be supplied to meet the desired level of service can be calculated.

Heuristic Procedures

To take account of the effect of traffic loading on travel times and traffic equilibria, several heuristic calculation procedures were developed. One heuristic proceeds incrementally. The traffic to be assigned is divided into parts (usually 4). Assign the first part of the traffic. Compute new travel times and assign the next part of the traffic. The last step is repeated until all the traffic is assigned. The CATS used a variation on this; it assigned row by row in the O-D table.

The heuristic included in the FHWA collection of computer programs proceeds another way.

- 0. Start by loading all traffic using an all or nothing procedure.
- 1. Compute the resulting travel times and reassign traffic.
- 2. Now, begin to reassign using weights. Compute the weighted travel times in the previous two loadings and use those for the next assignment. The latest iteration gets a weight of 0.25 and the previous gets a weight of 0.75.
- 3. Continue.

These procedures seem to work "pretty well," but they are not exact.

Frank-wolfe Algorithm

Dafermos (1968) applied the Frank-Wolfe algorithm (1956, Florian 1976), which can be used to deal with the traffic equilibrium problem. Suppose we are considering a highway network. For each link there is a function stating the relationship between resistance and volume of traffic. The Bureau of Public Roads (BPR) developed a link (arc) congestion (or volume-delay, or link performance) function, which we will term $S_a(v_a)$

$$S_a(v_a) = t_a\left(1 + 0.15\left(\frac{v_a}{c_a}\right)^4\right)$$

- t_a = free flow travel time on link *a* per unit of time
- v_a = volume of traffic on link *a* per unit of time (somewhat more accurately: flow attempting to use link *a*).
- c_a = capacity of link *a* per unit of time
- $S_a(v_a)$ is the average travel time for a vehicle on link *a*

There are other congestion functions. The CATS has long used a function different from that used by the BPR, but there seems to be little difference between results when the CATS and BPR functions are compared.

Equilibrium Assignment

To assign traffic to paths and links we have to have rules, and there are the well-known Wardrop equilibrium (1952) conditions. The essence of these is that travelers will strive to find the shortest (least resistance) path from origin to destination, and network equilibrium occurs when no traveler can decrease travel effort by shifting to a new path. These are termed user optimal conditions, for no user will gain from changing travel paths once the system is in equilibrium.

The user optimum equilibrium can be found by solving the following nonlinear programming problem

$$\min \sum_a \int_0^{v_a} S_a(x)dx$$

subject to:

$$v_a = \sum_i \sum_j \sum_r \alpha_{ij}^{ar} x_{ij}^r$$

$$\sum_r x_{ij}^r = T_{ij}$$

$$v_a \geq 0,\ x_{ij}^r \geq 0$$

where x_{ij}^r is the number of vehicles on path *r* from origin *i* to destination *j*. So constraint (2) says that all travel must take place $-i = 1 \ldots n; j = 1 \ldots n$

$\alpha_{ij}^{ar} = 1$ if link a is on path r from i to j ; zero otherwise. So constraint (1) sums traffic on each link. There is a constraint for each link on the network. Constraint (3) assures no negative traffic.

Example

An example from Eash, Janson, and Boyce (1979) will illustrate the solution to the nonlinear program problem. There are two links from node 1 to node 2, and there is a resistance function for each link. Areas under the curves in Figure 2 correspond to the integration from 0 to *a* in equation 1, they sum to 220,674. Note that the function for link *b* is plotted in the reverse direction.

$$S_a = 15\left(1 + 0.15\left(\frac{v_a}{1000}\right)^4\right)$$

$$S_b = 20\left(1 + 0.15\left(\frac{v_b}{3000}\right)^4\right)$$

$$v_a + v_b = 8000$$

Figure 1: Two Route Network

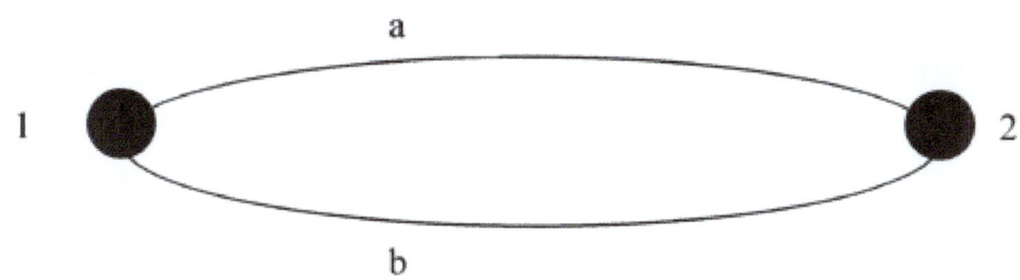

Figure 2: Graphical Solution to the Equilibrium Assignment Problem

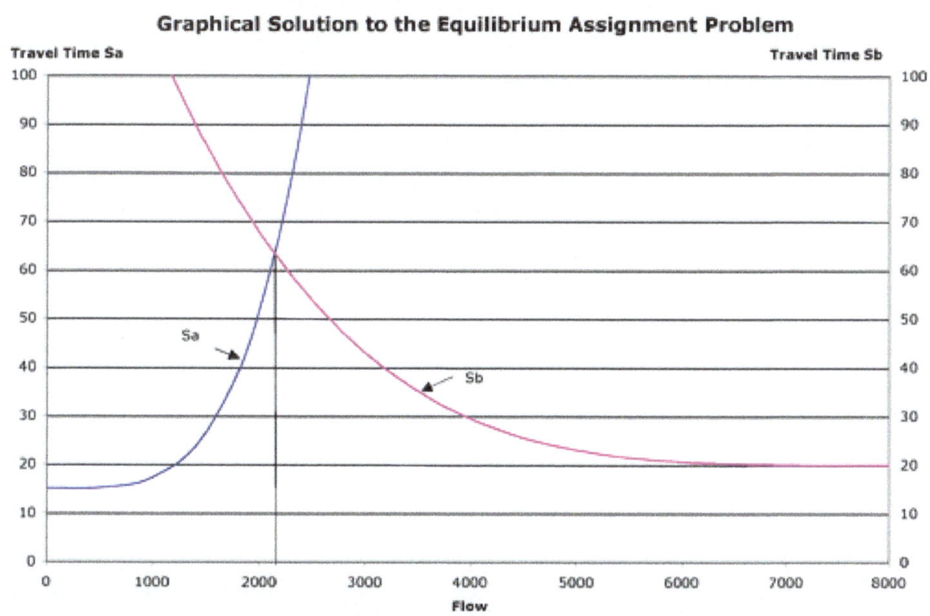

Figure 3: Allocation of Vehicles not Satisfying the Equilibrium Condition

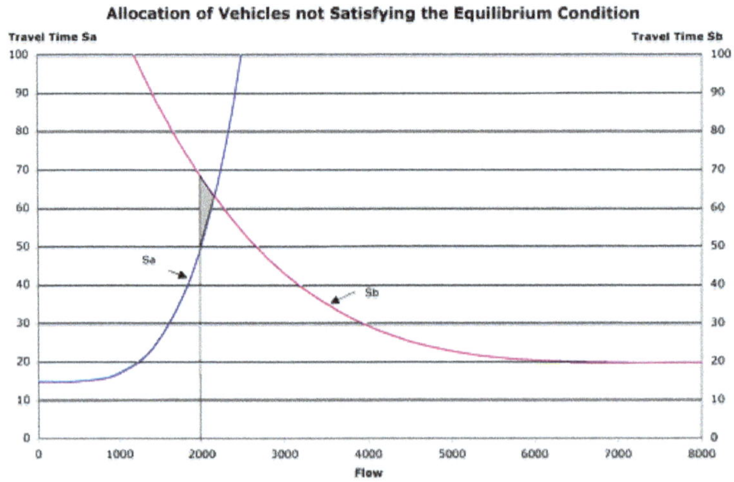

At equilibrium there are 2,152 vehicles on link *a* and 5847 on link *b*. Travel time is the same on each route: about 63.

Figure 3 illustrates an allocation of vehicles that is not consistent with the equilibrium solution. The curves are unchanged. But with the new allocation of vehicles to routes the shaded area has to be included in the solution, so the Figure 3 solution is larger than the solution in Figure 2 by the area of the shaded area.

Transit Assignment

There are also methods that have been developed to assign passengers to transit vehicles.

Integrating Travel Choices

The urban transportation planning model evolved as a set of steps to be followed, and models evolved for use in each step. Sometimes there were steps within steps, as was the case for the first statement of the Lowry model. In some cases, it has been noted that steps can be integrated. More generally, the steps abstract from decisions that may be made simultaneously, and it would be desirable to better replicate that in the analysis.

Disaggregate demand models were first developed to treat the mode choice problem. That problem assumes that one has decided to take a trip, where that trip will go, and at what time the trip will be made. They have been used to treat the implied broader context. Typically, a nested model will be developed, say, starting with the probability of a trip being made, then examining the choice among places, and then mode choice. The time of travel is a bit harder to treat.

Wilson's doubly constrained entropy model has been the point of departure for efforts at the aggregate level. That model contains the constraint

$$t_{ij}c_{ij} = C$$

where the c_{ij} are the link travel costs, t_{ij} refers to traffic on a link, and C is a resource constraint to be sized when fitting the model with data. Instead of using that form of the constraint, the monotonically increasing resistance function used in traffic assignment can be used. The result determines zone-to-zone movements and assigns traffic to networks, and that makes much sense from the way one would imagine the system works – zone-to-zone traffic depends on the resistance occasioned by congestion.

Alternatively, the link resistance function may be included in the objective function (and the total cost function eliminated from the constraints).

A generalized disaggregate choice approach has evolved as has a generalized aggregate approach. The large question is that of the relations between them. When we use a macro model, we would like to know the disaggregate behavior it represents. If we are doing a micro analysis, we would like to know the aggregate implications of the analysis.

Wilson derives a gravity-like model with weighted parameters that say something about the attractiveness of origins and destinations. Without too much math we can write probability of choice statements based on attractiveness, and these take a form similar to some varieties of disaggregate demand models.

Integrating Travel Demand with Route Assignment

It has long been recognized that travel demand is influenced by network supply. The example of a new bridge opening where none was before inducing additional traffic has been noted for centuries. Much research has gone into developing methods for allowing the forecasting system to directly account for this phenomenon. Evans (1974) published a doctoral dissertation on a mathematically rigorous combination of the gravity distribution model with the equilibrium assignment model. The earliest citation of this integration is the work of Irwin and Von Cube, as related by Florian et al. (1975), who comment on the work of Evans:

"The work of Evans resembles somewhat the algorithms developed by Irwin and Von Cube ["Capacity Restraint in Multi-Travel Mode Assignment Programs" H.R.B. Bulletin 347 (1962)] for a transportation study of Toronto, Canada. Their work allows for feedback between congested assignment and trip distribution, although they apply sequential procedures. Starting from an initial solution of the distribution problem, the interzonal trips are assigned to the initial shortest routes. For successive iterations, new shortest routes are computed, and their lengths are used as access times for input the distribution model. The new interzonal flows are then assigned in some proportion to the routes already found. The procedure is stopped when the interzonal times for successive iteration are quasi-equal."

Florian et al. proposed a somewhat different method for solving the combined distribution assignment, applying directly the Frank-Wolfe algorithm. Boyce et al. (1988) summarize the research on Network Equilibrium Problems, including the assignment with elastic demand.

Discussion

A three link problem can not be solved graphically, and most transportation network problems involve a large numbers of nodes and links. Eash et al., for instance, studied the road net on DuPage County where there were about 30,000 one-way links and 9,500 nodes. Because problems are large, an algorithm is needed to solve the assignment problem, and the Frank-Wolfe algorithm (modified a bit since first published) is used. Start with an all or nothing assignment, and then follow the rule developed by Frank-Wolfe to iterate toward the minimum value of the objective function. (The algorithm applies successive feasible solutions to achieve convergence to the optimal solution. It uses an efficient search procedure to move the calculation rapidly toward the optimal solution.) Travel times correspond to the dual variables in this programming problem.

It is interesting that the Frank-Wolfe algorithm was available in 1956. Its application was developed in 1968, and it took almost another two decades before the first equilibrium assignment algorithm was embedded in commonly used transportation planning software (Emme and Emme/2, developed by Florian and others in Montreal). We would not want to draw any general conclusion from the slow application observation, mainly because we can find counter examples about the pace and pattern of technique development. For example, the simplex method for the solution of linear programming problems was worked out and widely applied prior to the development of much of programming theory.

The problem statement and algorithm have general applications across civil engineering -- hydraulics, structures, and construction.

6

Traffic Engineering: An Overview

Traffic engineering is a branch of civil engineering; this subject focuses on the techniques used to achieve safe and efficient methods for road safety. This subject is closely related to transport engineering, pavement engineering, highway engineering and urban planning. This section helps the readers in developing an in-depth understanding of all the aspects of traffic engineering.

Traffic Engineering (Transportation)

Traffic engineering is a branch of civil engineering that uses engineering techniques to achieve the safe and efficient movement of people and goods on roadways. It focuses mainly on research for safe and efficient traffic flow, such as road geometry, sidewalks and crosswalks, cycling infrastructure, traffic signs, road surface markings and traffic lights. Traffic engineering deals with the functional part of transportation system, except the infrastructures provided.

Traffic engineering is closely associated with other disciplines:

- Transport engineering
- Pavement engineering
- Bicycle transportation engineering
- Highway engineering
- Transportation planning
- Urban planning
- Human factors engineering.

Typical traffic engineering projects involve designing traffic control device installations and modifications, including traffic signals, signs, and pavement markings. However, traffic engineers also consider traffic safety by investigating locations with high crash rates and developing countermeasures to reduce crashes. Traffic flow management can be short-term (preparing construction traffic control plans, including detour plans for pedestrian and vehicular traffic) or long-term (estimating the impacts of proposed commercial developments on traffic patterns). Increasingly, traffic problems are being addressed by developing systems for intelligent transportation systems, often in conjunction with other engineering disciplines, such as computer engineering and electrical engineering.

Traffic Systems

Traditionally, road improvements have consisted mainly of building additional infrastructure. However, dynamic elements are now being introduced into road traffic management. Dynamic elements have long been used in rail transport. These include sensors to measure traffic flows and automatic, interconnected, guidance systems to manage traffic (for example, traffic signs which open a lane in different directions depending on the time of day). Also, traffic flow and speed sensors are used to detect problems and alert operators, so that the cause of the congestion can be determined, and measures can be taken to minimize delays. These systems are collectively called intelligent transportation systems.

Lane Flow Equation

A ramp meter limits the rate at which vehicles can enter the freeway

The relationship between lane flow (Q, vehicles per hour), maximum speed (V, kilometers per hour) and density (K, vehicles per kilometer) is

$$Q = KV$$

Observation on limited access facilities suggests that up to a maximum flow, speed does not decline while density increases. However, above a critical threshold, increased density reduces speed. Additionally, beyond a further threshold, increased density reduces flow as well.

Therefore, speeds and lane flows at bottlenecks can be kept high during peak periods by managing traffic density using devices that limit the rate at which vehicles can enter the highway. Ramp meters, signals on entrance ramps that control the rate at which vehicles are allowed to enter the mainline facility, provide this function (at the expense of increased delay for those waiting at the ramps).

Highway Safety

Highway safety engineering is a branch of traffic engineering that deals with reducing the frequency and severity of crashes. It uses physics and vehicle dynamics, as well as road user psychology and human factors engineering, to reduce the influence of factors that contribute to crashes.

A typical traffic safety investigation follows these steps

1. Identify and prioritize investigation locations. Locations are selected by looking for sites with higher than average crash rates, and to address citizen complaints.

2. Gather data. This includes obtaining police reports of crashes, observing road user behavior, and collecting information on traffic signs, road surface markings, traffic lights and road geometry.

3. Analyze data. Look for collisions patterns or road conditions that may be contributing to the problem.

4. Identify possible countermeasures to reduce the severity or frequency of crashes.

• Evaluate cost/benefit ratios of the alternatives

• Consider whether a proposed improvement will solve the problem, or cause "crash migration." For example, preventing left turns at one intersection may eliminate left turn crashes at that location, only to increase them a block away.

• Are any disadvantages of proposed improvements likely to be worse than the problem you are trying to solve?

5. Implement improvements.

6. Evaluate results. Usually, this occurs some time after the implementation. Have the severity and frequency of crashes been reduced to an acceptable level? If not, return to step 2.

Traffic Sign

Traffic signs or road signs are signs erected at the side of or above roads to give instructions or provide information to road users. The earliest signs were simple wooden or stone milestones. Later, signs with directional arms were introduced, for example, the fingerposts in the United Kingdom and their wooden counterparts in Saxony.

With traffic volumes increasing since the 1930s, many countries have adopted pictorial signs or otherwise simplified and standardized their signs to overcome language barriers, and enhance traffic safety. Such pictorial signs use symbols (often silhouettes) in place of words and are usually based on international protocols. Such signs were first developed in Europe, and have been adopted by most countries to varying degrees.

Categories

Traffic signs can be grouped into several types. For example, Annexe 1 of the Vienna Convention on Road Signs and Signals (1968), which on 30 June 2004 had 52 signatory countries, defines eight categories of signs:

- A. Danger warning signs
- B. Priority signs
- C. Prohibitory or restrictive signs
- D. Mandatory signs

- E. Special regulation signs
- F. Information, facilities, or service signs
- G. Direction, position, or indication signs
- H. Additional panels

In the United States, Canada, Ireland, Australia, and New Zealand signs are categorized as follows:

- Regulatory signs
- Warning signs
- Guide signs
    - Street name signs
    - Route marker signs
    - Expressway signs
    - Freeway signs
    - Welcome signs
    - Informational signs
    - Recreation and cultural interest signs
- Emergency management (civil defense) signs
- Temporary traffic control (construction or work zone) signs
- School signs
- Railroad and light rail signs
- Bicycle signs

Sign warning of cattle crossing in a rural road of Madeira island, Portugal

In the United States, the categories, placement, and graphic standards for traffic signs and pavement markings are legally defined in the Federal Highway Administration's *Manual on Uniform Traffic Control Devices* as the standard.

An rather informal distinction among the directional signs is the one between advance directional signs, interchange directional signs, and reassurance signs. Advance directional signs appear at a certain distance from the interchange, giving information for each direction. A number of countries do not give information for the road ahead (so-called "pull-through" signs), and only for the directions left and right. Advance directional signs enable drivers to take precautions for the exit (e.g., switch lanes, double check whether this is the correct exit, slow down). They often do not appear on lesser roads, but are normally posted on expressways and motorways, as drivers would be missing exits without them. While each nation has its own system, the first approach sign for a motorway exit is mostly placed at least 1000 m from the actual interchange. After that sign, one or two additional advance directional signs typically follow before the actual interchange itself.

History

Roman milestone

17th century traffic sign in Salvador street, Lisbon, Portugal stating which traffic should back up to give way: *Year of 1686. His Majesty commands all coaches, carriages and litters coming from Salvador's entrance to back up to the same part*

Speed camera sign used in Canada, Cyprus, the Czech Republic, Estonia, Finland, Georgia, Hong Kong, Iceland, Iran, Ireland, Latvia, Malta, Poland, Ukraine, the United Arab Emirates, and the United Kingdom

The earliest road signs were milestones, giving distance or direction; for example, the Romans erected stone columns throughout their empire giving the distance to Rome. In the Middle Ages, multidirectional signs at intersections became common, giving directions to cities and towns.

In 1686, the first known Traffic Regulation Act in Europe is established by King Peter II of Portugal. This act foresees the placement of priority signs in the narrowest streets of Lisbon, stating which traffic should back up to give way. One of these signs still exists at Salvador street, in the neighborhood of Alfama.

The first modern road signs erected on a wide scale were designed for riders of high or "ordinary" bicycles in the late 1870s and early 1880s. These machines were fast, silent and their nature made them difficult to control, moreover their riders travelled considerable distances and often preferred to tour on unfamiliar roads. For such riders, cycling organizations began to erect signs that warned of potential hazards ahead (particularly steep hills), rather than merely giving distance or directions to places, thereby contributing the sign type that defines "modern" traffic signs.

The development of automobiles encouraged more complex signage systems using more than just text-based notices. One of the first modern-day road sign systems was devised by the Italian Touring Club in 1895. By 1900, a Congress of the International League of Touring Organizations in Paris was considering proposals for standardization of road signage. In 1903 the British government introduced four "national" signs based on shape, but the basic patterns of most traffic signs were set at the 1908 International Road Congress in Paris. In 1909, nine European governments agreed on the use of four pictorial symbols, indicating "bump", "curve", "intersection", and "grade-level railroad crossing". The intensive work on international road signs that took place between 1926 and 1949 eventually led to the development of the European road sign system. Both Britain and the United States developed their own road signage systems, both of which were adopted or modified by many other nations in their respective spheres of influence. The UK adopted a version of the European road signs in 1964 and, over past decades, North American signage began using some symbols and graphics mixed in with English.

Over the years, change was gradual. Pre-industrial signs were stone or wood, but with the development of Darby's method of smelting iron using coke, painted cast iron became favoured in the late 18th and 19th centuries. Cast iron continued to be used until the mid-20th century, but it was gradually displaced by aluminium or other materials and processes, such as vitreous enamelled and/or pressed malleable iron, or (later) steel. Since 1945 most signs have been made from sheet aluminium with adhesive plastic coatings, these are normally retroreflective for nighttime and low-light visibility. Before the development of reflective plastics, reflectivity was provided by glass reflectors set into the lettering and symbols.

New generations of traffic signs based on electronic displays can also change their text (or, in some countries, symbols) to provide for "intelligent control" linked to automated traffic sensors or remote manual input. In over 20 countries, real-time Traffic Message Channel incident warnings are conveyed directly to vehicle navigation systems using inaudible signals carried via FM radio, 3G cellular data and satellite broadcasts. Finally, cars can pay tolls and trucks pass safety screening checks using video numberplate scanning, or RFID transponders in windshields linked to antennae over the road, in support on-board signalling, toll collection and travel time monitoring.

Yet another "medium" for transferring information ordinarily associated with visible signs is RIAS (Remote Infrared Audible Signage), e.g., "talking signs" for print-handicapped (including blind/low-vision/illiterate) people. These are infra-red transmitters serving the same purpose as the usual graphic signs when received by an appropriate device such as a hand-held receiver or one built into a cell phone.

North America, Australia and New Zealand

handicap sign

One of Catskill Park's distinctive brown town signs with yellow text, showing the hamlet of Pine Hill

Color Schemes

The North American, Australian and New Zealand colors normally have these meanings:

- red with white for stop signs, yield, and forbidden actions (such as No Parking)
- green with white letters for informational signs, such as directions, distances, and places
- brown with white letters for signs to parks, historic sites, ski areas, forests, and campgrounds
- blue with white symbols (or business logos) for rest areas, food, gasoline, hospitals, lodging, and other services
- white with black (or red) letters for regulatory signs, such as speed limits (or parking)
- yellow with black letters and symbols for warning signs, such as curves and school zones
- orange with black letters for temporary traffic control zones and detours associated with road construction
- purple for "lanes restricted to use only by vehicles with registered electronic toll collection (ETC) accounts", such as EZPass.
- black with white letters or arrows for lane use.

The US *Manual on Uniform Traffic Control Devices* prescribes four other colors:

- fluorescent yellow-green with black symbols for school zone, school bus stop, pedestrian, playground, and bicycle warning signs
- fluorescent pink with black letters and symbols for incident management signs
- coral and light blue, which are unassigned but reserved for potential future use.

Regulatory signs are also sometimes seen with white letters on red or black signs. In Quebec, blue is often used for public services such as rest areas; many black-on-yellow signs are red-on-white instead.

Many US states and Canadian provinces now use fluorescent orange for construction signs.

Highway Symbols and Markers

Rural highway sign, Saskatchewan.

Every state and province has different markers for its own highways, but use standard ones for all federal highways. Many special highways – such as the Queen Elizabeth Way, Trans-Canada Highway, and various auto trails in the U.S. – have used unique signs. Counties in the US sometimes use a pentagonal blue sign with yellow letters for numbered county roads, though the use is inconsistent even within states.

Units

Distances on traffic signs generally follow the measurement system in use locally: that is to say, the metric system in all countries of the world except Burma, Liberia, the United Kingdom, and the United States – although the metric system is used in the UK for all purposes other than the display of road distances and the defining of speed limits, and in the US the Federal Department of Transportation has developed (very rarely used) metric standards for all signs.

Languages

Multilingual road signs in Mistissini, Quebec in Cree, English and French.

Where signs use a language, the recognized language/s of the area is normally used. Signs in most of the US, Canada, Australia, and New Zealand are in English. Quebec uses French, while New Brunswick and the Jacques-Cartier and Champlain bridges, in Montreal (as well as some parts in the West Island), use both English and French, and a number of other provinces and states, such as Ontario, Manitoba, and Vermont use bilingual French–English signs in certain localities. Puerto Rico (a US territory), Mexico, and Spain use Spanish. Within a few miles of the US–Mexico border, road signs are often in English and Spanish in places like San Diego, Yuma, and El Paso. Indigenous languages, mainly Nahuatl as well as some Mayan languages, have been used as well.

Typefaces

The typefaces predominantly used on signs in the US and Canada are the FHWA alphabet series (Series B through Series F and Series E Modified). Details of letter shape and spacing for these alphabet series are given in "Standard Alphabets for Traffic Control Devices", first published by the Bureau of Public Roads (BPR) in 1945 and subsequently updated by the Federal Highway Administration (FHWA). It is now part of Standard Highway Signs (SHS), the companion volume to the MUTCD which gives full design details for signfaces.

Initially, all of the alphabet series consisted of uppercase letters and digits only, although lowercase extensions were provided for each alphabet series in a 2002 revision of SHS. Series B through Series F evolved from identically named alphabet series which were introduced in 1927.

Straight-stroke letters in the 1927 series were substantially similar to their modern equivalents, but unrounded glyphs were used for letters such as B, C, D, etc., to permit more uniform fabrication of signs by illiterate painters. Various state highway departments and the federal BPR experimented with rounded versions of these letters in the following two decades.

The modern, rounded alphabet series were finally standardized in 1945 after rounded versions of some letters (with widths loosely appropriate for Series C or D) were specified as an option in the 1935 MUTCD and draft versions of the new typefaces had been used in 1942 for guide signs on the newly constructed Pentagon road network.

The mixed-case alphabet now called Series E Modified, which is the standard for destination legend on freeway guide signs, originally existed in two parts: an all-uppercase Series E Modified, which was essentially similar to Series E, except for a larger stroke width, and a lowercase-only alphabet. Both parts were developed by the California Division of Highways (now Caltrans) for use on freeways in 1948–1950.

Initially, the Division used all-uppercase Series E Modified for button-reflectorized letters on ground-mounted signs and mixed-case legend (lowercase letters with Series D capitals) for externally illuminated overhead guide signs. Several Eastern turnpike authorities blended all-uppercase Series E Modified with the lowercase alphabet for destination legends on their guide signs.

Eventually, this combination was accepted for destination legend in the first manual for signing Interstate highways, which was published in 1958 by the American Association of State Highway Officials and adopted as the national standard by the BPR.

Uses of Non-FHWA Typefaces

Some traffic signs, such as the left-turn prohibition sign hanging from this gantry, are lit for better visibility, particularly at night or in inclement weather.

The US National Park Service uses NPS Rawlinson Roadway, a serif typeface, for guide signage; it typically appears on a brown background. Rawlinson has replaced Clarendon as the official NPS typeface, but some states still use Clarendon for recreational signage.

Georgia, in the past, used uppercase Series D with a custom lowercase alphabet on its freeway guide signs; the most distinctive feature of this typeface is the lack of a dot on lowercase *i* and *j*. More recent installations appear to include the dots.

The Clearview typeface, developed by US researchers to provide improved legibility, is permitted for light legend on dark backgrounds under FHWA interim approval. Clearview has seen widespread use by state departments of transportation in Arkansas, Arizona, Illinois, Kentucky, Maryland, Michigan, Ohio, Pennsylvania, Texas, and Virginia. The Kansas Turnpike Authority has also introduced Clearview typeface to some of its newer guide signs along the Kansas Turnpike, but the state of Kansas continues to use the FHWA typefaces for signage on its non-tolled Interstates and freeways.

In Canada, the Ministry of Transportation for the Province of British Columbia specifies Clearview for use on its highway guide signs, and its usage has shown up in Ontario on the Don Valley Parkway and Gardiner Expressway in Toronto and on new 400-series highway installations in Hamilton, Halton and Niagara, as well as street signs in various parts of the province. The font is also being used on newer signs in Alberta, Manitoba, and Quebec.

A new Clearview typeface sign beside an old FHWA typeface, Quebec

Moose crossing warning with kill-counter, Kenai Peninsula of Alaska. Prior to widespread human settlement, this area was part of the Kenai National Moose Range.

It is common for local governments, airport authorities, and contractors to fabricate traffic signs using typefaces other than the FHWA series; Helvetica, Futura and Arial are common choices.

New Zealand

New Zealand road signs are generally influenced both by American and European practices.

Warning signs are diamond-shaped with a yellow background for permanent warnings, and an orange background for temporary warnings. They are somewhat more pictorial than their American counterparts. This is also true for Canadian signage.

Regulatory signs also follow European practice, with a white circle with a red border indicating prohibitive actions, and a blue circle indicating mandatory actions. White rectangular signs with a red border indicate lane usage directions. Information and direction signs are rectangular, with a green background indicating a state highway, a blue background for all other roads and all services (except in some, where directional signage is white), and a brown background for tourist attractions.

Before 1987, most road signs had black backgrounds – diamonds indicated warnings, and rectangles indicated regulatory actions (with the exception of the Give Way sign (an inverted trapezium), and Stop sign and speed limit signs (which were the same as today)). Information signs were yellow, and direction signage was green on motorways and black everywhere else.

Europe

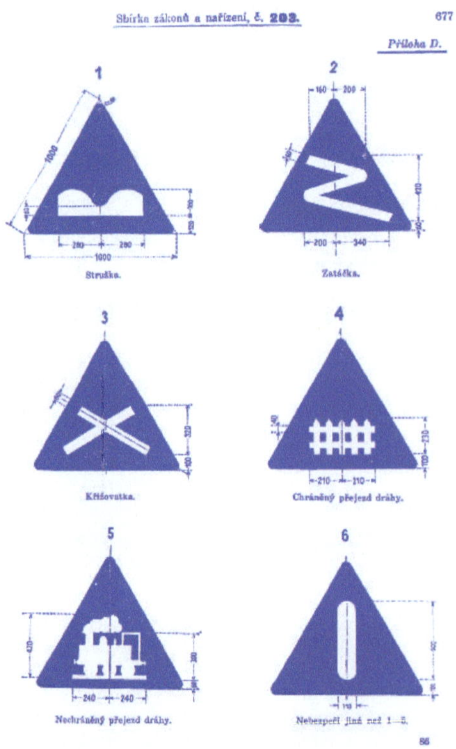

The first road signs established in Czechoslovakia on 1 November 1935: six blue-white danger warning signs. They were later supplanted with red-white-black signs.

Keep right, Portugal.

Road sign in Beussent, France - entrance to built up area with an implied 50 km/h speed limit.

Since the signing of the 1931 Geneva Convention concerning the Unification of Road Signals by a number of countries that the standardization of the traffic signs started in Europe. The 1931 Convention rules were developed in the 1949 Geneva Protocol on Road Signs and Signals.

In 1968, the European countries signed the Vienna Convention on Road Traffic treaty, with the aim of standardizing traffic regulations in participating countries in order to facilitate international road traffic and to increase road safety. Part of the treaty was the Vienna Convention on Road Signs and Signals, which defined the traffic signs and signals. As a result, in Western Europe the traffic signs are well standardized, although there are still some country-specific exceptions, mostly dating from the pre-1968 era.

The principle of the European traffic sign standard is that certain shapes and colours are to be used with consistent meanings:

- Triangular signs (black symbols on a white or yellow background) warn of dangers. The Vienna Convention additionally allows an alternative shape for such signs, namely a right-angled diamond – although in Europe this shape is regularly used only in the Republic of Ireland.
- Regulatory signs are round: those indicating a prohibition or limit are black on white (or yellow) with a red border; those giving a mandatory order are white on blue.
- Informational and various other secondary signs are of rectangular shape.

- The animals which may be depicted on warning signs include cattle, deer, ducks, elk, frogs, horses, sheep, monkeys (in Gibraltar), and polar bears (on Svalbard). The Convention allows any animal image to be used.

Directional signs ("guide signs" in American parlance) have not been harmonized under the Convention, at least not on ordinary roads. As a result, there are substantial differences in directional signage throughout Europe. Differences apply to the choice of typeface, arrows and, most notably, colours. The convention does, however, specify that the type of directional signage used should, for each country, distinguish limited-access roads ("motorways") from ordinary, all-purpose roads.

Directional signage on motorways uses:

- white-on-green in, for example, Albania, Bosnia and Herzegovina, Bulgaria, Croatia, Cyprus, the Czech Republic, Denmark, Finland, Greece, Italy, the Republic of Macedonia, Romania, Serbia, Slovakia, Slovenia, Sweden, Switzerland, and Turkey
- white-on-blue in, for example, Austria, Belarus, Belgium, France, Germany, Hungary, Luxembourg, the Netherlands, Norway, Poland, Portugal, the Republic of Ireland, Spain, and the United Kingdom.

Differences are greater for non-motorways:

- White-on-blue in Italy, Switzerland, Sweden, the Czech Republic, Greece, Cyprus, Slovakia, Bulgaria, Romania, Latvia, Estonia, Finland, Turkey, and the Netherlands (in this case the same as on motorways).
- White-on-green in the United Kingdom, the Republic of Ireland, Poland, France, and Portugal (only on the few primary roads that still have not been transformed in motorways),
- Black-on-yellow in Bosnia and Herzegovina, Germany, Luxembourg, Norway, Slovenia, Serbia, and Croatia.
- Red-on-white in Denmark (though white-on-blue on motorway exits and all overhead gantries)
- Black-on-white in Spain.

The black-on-white signposting of secondary roads distinguishes them from primary roads in Finland, France, Portugal, the Republic of Ireland, Switzerland, and the United Kingdom. In Germany, Italy, Romania, and Sweden black-on-white indicates urban-only roads or urban destinations.

The signposting of road numbers also differs greatly, except that European route numbers, if displayed, are always indicated using white characters on a green rectangle. European route numbers are, however, not signed at all in the United Kingdom.

The Convention recommends that certain signs – such as "STOP", "ZONE", etc. – be in English; however, use of the local language is also permitted. If a language uses non-Latin characters, a Latin-script transliteration of the names of cities and other important places should also be given. Road signs in the Republic of Ireland are bilingual, using Irish and English. Wales similarly uses bilingual Welsh–English signs, while some parts of Scotland have bilingual Scottish Gaelic–English signs. Finland also uses bilingual signs, in Finnish and Swedish. Signs in Belgium are in French, Dutch, or German depending on the region. In the Brussels Capital Region, road signs are in both French and Dutch. Signs in Switzerland are in French, German, or Italian depending on the canton.

European countries – with the notable exception of the United Kingdom, where distances and lengths are indicated in miles, yards, feet, and inches, and speed limits are expressed in miles per hour – use the metric system on road signs.

For countries driving on the left, the convention stipulates that the traffic signs should be mirror images of those used in countries driving on the right. This practice, however, is not systematically followed in the four European countries driving on the left, Cyprus, Malta, the Republic of Ireland, and the United Kingdom. The convention permits the use of two background colours for danger and prohibition signs: white or yellow. Most countries use white, with a few – such as Finland, Iceland, Poland, and Sweden – opting for yellow as this tends to improve the winter-time visibility of signs in areas where snow is prevalent. In some countries, such as France, white is the normal background colour for such signs, but yellow is used for temporary signage (as, for example, at road works).

European traffic signs have been designed with the principles of heraldry in mind; i.e., the sign must be clear and able to be resolved at a glance. Most traffic signs conform to heraldic tincture rules, and use symbols rather than written texts for better semiotic clarity.

Croatia

Croatian road signs follow the Vienna convention (SFR Yugoslavia was the original signatory for Croatia, which is now a contracting party itself). The most common signs are:

- Yellow and black signs *for direction.*
- Blue and white signs *for information.*
- White-on-green signs are used on the highways.

In the first years following Croatia's independence, its traffic signs were the same as in the rest of the former Yugoslavia. In the early 2000s, replacement of the yellow background of warning signs began, and new signs now use a white background.

The signage typeface is SNV, as with the other countries of the former Yugoslavia.

Iceland

Road signs in Iceland mainly follow the Vienna Convention, but use a variant of the colour scheme and minor design changes similar to the signs in Sweden.

Ireland

Until the partition of Ireland in 1922 and the independence of the Irish Free State (now the Republic of Ireland), British standards applied across the island. In 1926 road sign standards similar to those used in the UK at the time were adopted. Law requires that the signs be written in both Irish and English.

Advance directional sign in for a roundabout in Ireland. The green background indicates that this sign is on a national road, with the blue patches left and right indicating a motorway (with symbol) and the white patches indicating a regional road or local road.

Irish warning sign

In 1956, road signs in the Republic were changed to markedly differ from the UK standard, with the adoption of US-style "diamond" signs for many road hazard warnings (junctions, bends, railway crossings, traffic lights). Some domestic signs were also invented, such as the keep-left sign (a black curved arrow pointing to the upper-left, although some are similar to the European "white arrow on blue disk" signs), while some other signs are not widely adopted outside Ireland, such as the no-entry sign (a black arrow pointing ahead in a white circle with a red slashed circumference).

Directional signage is still firmly based on the United Kingdom standard, however, with the basic design of directional signs remaining the same as the UK in most cases. The same colours are used for directional signs in Ireland as in the UK, and the UK Transport and Motorway fonts are used. However, signage in the Republic of Ireland is bilingual, with the Irish text in mixed case italics, while the English text is in all upper-case.

In January 2005 Ireland adopted metric speed limits. Around 35,000 existing signs were replaced and a further 23,000 new signs erected bearing the speed limit in kilometres per hour. To avoid confusion with the old signs, each speed limit sign now has "km/h" beneath the numerals. Also, since the adoption of signs based on the *Warboys Committee* standard in 1977, Irish directional signs have used the metric system, however, unlike with the later speed limit change over, there was no effort made to change the existing signage, and as of 2007 many finger posts still remain on rural roads with distances in miles, although the numbers continue to decline as roads are improved.

In late 2007 Ireland began an extensive programme of sign and post replacement. Good examples are the M1 (Dublin–Dundalk) and the M50 (Dublin). While being mostly the same as the old signs, it is welcome as a lot of the signs were damaged/stained. About half of the new posts are now two medium posts with crosshatched metal posts in-between instead of one large pole to minimise the damage in case of a crash.

Latvia

Road signs in Latvia largely adhere to Vienna Convention guidelines. In detailed design they closely resemble the signs used in Germany.

Netherlands

Road signs in the Netherlands follow the Vienna Convention. Directional signs (which have not been harmonized under the Convention) always use blue as the background colour. The destinations on the sign are printed in white. If the destination is not a town (but an area within town or some other kind of attraction), that destination will be printed in black on a separate white background within the otherwise blue sign.

The Netherlands always signposts European road numbers where applicable (i.e., on the advance directional signs, the interchange direction signs and on the reassurance signs). Dutch national road numbers are placed on a rectangle, with motorways being signposted in white on a red rectangle (as an A*xx*) and primary roads in black on a yellow rectangle (as N*xx*). When a motorway changes to a primary road, its number remains the same, but the A is replaced by the N. So at a certain point the A2 becomes N2, and when it changes to a motorway again, it becomes A2 again.

Signs intended for bike-riders always go on white signs with red or green letters.

The Dutch typeface, known as ANWB-Ee, is based on the US typeface. A new font, named ANWB-Uu (also known as Redesign), has been developed in 1997 and appears on many recent Dutch signs. On the motorways however the typeface remains the ANWB-Ee or a similar typeface. The language of the signs is typically Dutch, even though bilingual signs may be used, when the information is relevant for tourists.

Norway

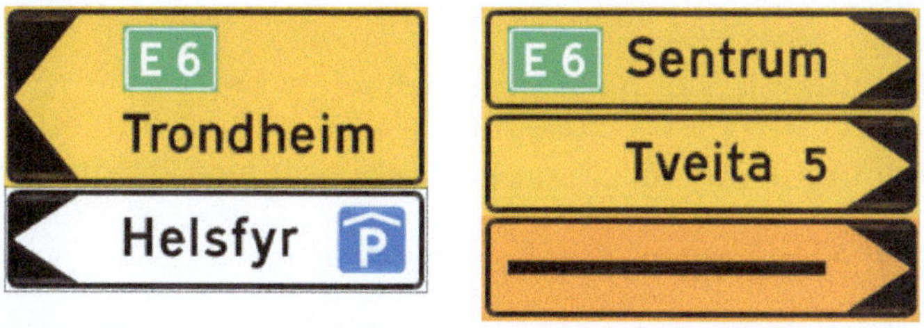

Upper left and right and middle right are standard directional signs. Lower left is for a commercial facility, and lower right is for a temporary detour.

Signs in Norway mostly follow the Vienna Convention, except the polar bear warning sign, which is a white bear on a black background and a red border. These are the directional signs:

- Signs for motorways are blue with white text
- Those for regular roads to towns and cities are yellow with black lettering
- Signs for industrial areas, commercial facilities etc. are white with black letters
- Signs for tourist attractions, national parks, museums etc. are brown with white lettering

The signs for road numbering are rectangular, and have this colour scheme:

- European routes (E6,E18,etc.)are green with white lettering
- National routes are also green with white lettering
- County owned roads are white with black lettering
- Municipality owned roads have the name of the road, instead of a number, and are white with black lettering

Sweden

Swedish elk warning sign

The road signs in Sweden mostly follow the Vienna Convention with a few adaptations, however, allowed within the convention:

- The background of warning signs is yellow
- Warning signs for elk and reindeer
- The background of direction signs is blue with white text
- The background of motorway direction signs is green with white text
- When applicable, the language is Swedish in Sweden.

The signage typeface Tratex is used exclusively in Sweden and is available as freeware.

United Kingdom

One of the more unusual UK road signs, at the Magic Roundabout in Swindon

Bilingual road sign in Wales

Traffic signing in the UK conforms broadly to European norms, though a number of signs are unique to Britain and direction signs omit European route numbers. The current sign system, introduced on 1 January 1965, was developed in the late 1950s and early 1960s by the Anderson Committee, which established the motorway signing system, and by the Worboys Committee, which reformed signing for existing all-purpose roads.

The UK remains the only European Union member nation and the only Commonwealth country to use non-metric (Imperial) measurements for distance and speed, although "authorised weight" signs have been in metric tonnes since 1981 and there is currently a dual-unit (metric first) option for height and width restriction signage, intended for use on safety grounds. On motorways kilometre signs are visible at intervals of 500 m indicating the distance from the start of the motorway.

Three colour schemes exist for direction signs:

- On motorways they are blue with white lettering
- On primary routes they are green with white lettering and yellow route numbers

- A non-primary route has white signs with black lettering
- A fourth colour scheme, black on yellow, is seen on temporary signs, for example marking a diversionary route avoiding a road closure.

Two typefaces are specified for British road signs. Transport "Medium" or Transport "Heavy" are used for all text on fixed permanent signs and most temporary signage, depending on the colour of the sign and associated text colour; dark text on a white background is normally set in "Heavy" so that it stands out better. However route numbers on motorway signs use a taller limited character set typeface called "Motorway".

Signs are generally bilingual in all parts of Wales (English/Welsh or Welsh/English), and similar signs are beginning to be seen in parts of the Scottish Highlands (English/Scottish Gaelic).

All signs and their associated regulations can be found in the Traffic Signs Regulations and General Directions, as updated by the TSRGD 2008 and TSRGD 2011 and complemented by the various chapters of the "Traffic Signs Manual".

Mexico, South and Central America

Speed bump sign in Belize.

Road signs in Mexico, Central America, and South America vary from country to country. For the most part, conventions in signage tend to resemble United States signage conventions more so than European and Asian conventions. For example, warning signs are typically diamond-shaped and yellow rather than triangular and white. Some variations include the "Parking" and "No Parking" signs, which contain either a letter *E* or *P*, depending on which word is used locally for "Parking" (Spanish *estacionamiento* or *parqueo*, Portuguese *estacionamento*), as well as the Stop sign, which usually reads "Pare" or "Alto". Notable exceptions include speed limit signs, which follow the European conventions, and the "No Entry" sign, often replaced with a crossed upwards arrow.

Colombia

Traffic signs in Colombia are classified into three categories:

- Warning signs
- Mandatory signs
- Information signs.

Warning signs are very similar to warning signs in United States. They are yellow diamond-shaped with a black symbol (the yellow colour is changed to an orange colour in areas under construction). In certain cases, the yellow colour is shifted to fluorescent yellow (in the School area sign and Chevron sign).

Mandatory signs are similar to European signs. They are circular with a red border, a white background and a black symbol. Stop sign and Yield sign are as European, except the word "Stop" is changed for "Pare" and the Yield sign has no letters, it is a red triangle with white centre.

Information signs have many shapes and colours. Principally they are blue with white symbols and in many cases these signs have an information letter below the symbol.

Africa

"T" junction road sign on a desert track in Niger.

Traffic Sign in Eritrea

Mauritius

Road signs in Mauritius are regulated by the Traffic Signs Regulations 1990; they are particularly modelled on the United Kingdom road signs since Mauritius is a former British colony. Mauritius has left-hand traffic.

Asia

Traffic sign demonstrating direction to Kuwait City, Kuwait

Road signs in Hong Kong

Warning signs for a camel are common signs In the region of Arabian Peninsula.

China

- Warning signs in China are triangular with a black border, yellow background and black symbol.
- Mandatory signs generally follow European conventions (circular with red border/blue circle) with some local variations.

Direction signs are:

- Green for expressways
- Brown for tourist attractions
- And blue for other roads.
- Occasionally black on white is used for directions to local facilities.

Hong Kong

Hong Kong's traffic signs follow the type of United Kingdom and are bi-lingual in English and Chinese (English on top, and Traditional Chinese characters at the bottom).

Iran

Road signs in Iran mainly follow the Vienna Convention. Signs are in Persian and English.

Israel

Road signs in Israel mainly follow the Vienna Convention, but have some variants.

Japan

Japanese stop sign with the word Tomare (止まれ), meaning Stop

Road signs in Japan are either controlled by local police authorities under Road Traffic Law (道路交通法 *Dōro Kōtsūhō*) or by other road-controlling entities including Ministry of Land, Infrastructure, Transport and Tourism, local municipalities, NEXCO (companies controlling expressways), under Road Law (道路法 *Dōrohō*). Most of the design of the road signs in Japan are similar to the signs on the Vienna Convention, except for some significant variances, such as stop sign with a red downward triangle. The main signs are categorized into four meaning types:

- Guidance (white characters on blue in general – on green in expressways),
- Warning (black characters and symbols on yellow diamond),
- Regulation (red or blue circle, depending on prohibition or regulation),
- And instruction (mostly white characters or symbols on blue square).

Philippines

Road warning sign in Bokod, Benguet

Route marker sign for Asian Highway 26, as seen on EDSA and the Maharlika Highway.

Road signs in the Philippines are standardized in the *Road Signs and Pavement Markings Manual*, published by the Department of Public Works and Highways. Philippine road signage practice closely follow the Vienna Convention, but with local adaptations and some minor influences from the US MUTCD and Australian road signs. However, some road signs may differ by locale, and mostly diverge from the national standard. For example, the Metropolitan Manila Development Authority (MMDA) has used pink and light blue in its signage for which it has been heavily criticised.

Road signs in the Philippines are classified as:

- Regulatory signs
- Warning signs
- Guide signs
- Expressway signs
- Traffic instruction signs

Regulatory road signs – other than the stop and give way signs – are generally circular, with (for prohibitions) a black symbol on a white background within a red border, or (for mandatory instructions) a white symbol on a blue background. In some cases circular regulatory signs are placed on white rectangular panels together with text supplementing their meanings.

Most warning signs display a black symbol on a white background within a red-bordered equilateral triangle. Since 2012, however, a more visibly distinctive design (taken from that used for school signs in the US) has been adopted for pedestrian-related signs: these consist of a fluorescent yellow-green pentagon with black border and symbol. Additional panels may be placed below signs to supplement their meanings.

Guide signs are divided into directional signs, service area signs, route markers, and tourist-related signs, with influence from both American and Australian practice. Directional signs use a green background with white letters and arrows. Service area signs use a blue background with white letters, arrows, and symbols. Tourist-related signs use a brown background with white letters, arrows, and symbols. The route marker sign, excluding the AH26 route marker, is based on the Australian National Route marker, but reserved for future use.

Signs on expressways mostly take elements from Australian motorway/freeway signs. Exit signs, wrong way signs and start/end of expressway signs are very similar to Australian freeway signage. Traffic instruction signs are textual signs used to supplement warning and regulatory signs.

Singapore

Singapore's traffic signs closely follow British road sign conventions, although the government has introduced some changes to them.

Sri Lanka

Road signs in Sri Lanka are standardized road signs closely follow those used in the United Kingdom with certain distinctions, and a number of changes have introduced road signs that suit as per local road and system. Sri Lankan government announced by a gazette that aimed to get a face-lift and introduction of over 100 new road traffic signs.

Taiwan

Automatic Traffic Sign Recognition

Cars are beginning to feature cameras with automatic traffic sign recognition, beginning with the Opel Insignia. It mainly recognizes speed limits and no-overtaking areas.

Street Sign Theft

Street sign theft occurs when street signs are stolen, often to be used as decorations, but also sometimes to avoid obeying the law by claiming later the sign was not there. Although the theft often seems arbitrary, signs that are unusual or amusing tend to be stolen more frequently. Sometimes considered to be a prank by the perpetrators, the theft is often costly and inconveniencing for the municipality or agency that owns the sign. In the United States, each street sign generally costs between $100 and $500 to replace.

Popular culture can act as a catalyst to street sign theft. Bands such as The Beatles and Lynyrd Skynyrd have exacerbated street sign theft as their songs and albums include real place names including Penny Lane, Blue Jay Way, Abbey Road, and Brickyard Road.

A provocative street name or other name on the sign can also trigger thefts. Well-known examples include Ragged Ass Road in Yellowknife, Northwest Territories, Butt Hole Road in Conisbrough, Doncaster, England and Sodom off the B5429 in Denbighshire, Wales.

Traffic Light

Traffic lights, also known as traffic signals, traffic lamps, traffic semaphore, signal lights, stop lights, (in South African English) robots and (in technical parlance) traffic control signals, are signalling devices positioned at road intersections, pedestrian crossings, and other locations to control flows of traffic.

The world's first, manually operated gas-lit traffic signal was short lived. Installed in London in December 1868, it exploded less than a month later, injuring or killing its policeman operator. Traffic control started to seem necessary in the late 1890s and Earnest Sirrine from Chicago patented the first automated traffic control system in 1910. It used the words "STOP" and "PROCEED", though both words did not light up.

Traffic lights alternate the right of way accorded to users by displaying lights of a standard color (red, amber (yellow), and green) following a universal color code. In the typical sequence of color phases:

- The green light allows traffic to proceed in the direction denoted, if it is safe to do so and there is room on the other side of the intersection.
- The amber (yellow) light warns that the signal is about to change to red. In a number of countries – among them the United Kingdom – a phase during which red and yellow are displayed together indicates that the signal is about to change to green. Actions required by drivers on a yellow light vary, with some jurisdictions requiring drivers to stop if it is safe to do so, and others allowing drivers to go through the intersection if safe to do so.
- A flashing Amber indication is a warning signal. In the United Kingdom, a flashing amber light is used only at pelican crossings, in place of the combined red–amber signal, and indicates that drivers may pass if no pedestrians are on the crossing.
- The red signal prohibits any traffic from proceeding.
- A flashing red indication is treated as a stop sign.

In some countries traffic signals will go into a flashing mode if the controller detects a problem, such as a program that tries to display green lights to conflicting traffic. The signal may display flashing yellow to the main road and flashing red to the side road, or flashing red in all directions. Flashing operation can also be used during times of day when traffic is light, such as late at night.

History

A traffic light in Stockholm in 1953.

The installation of a traffic signal in San Diego in December 1940

Marshalite traffic signal. Formerly fitted in various intersections in Melbourne, Australia, indicating how much time remained before a signal change.

On 9 December 1868, the first, non-electric, gas-lit traffic lights were installed outside the Houses of Parliament in London to control the traffic in Bridge Street, Great George Street and Parliament Street. They were promoted by the railway engineer J. P. Knight and constructed by the railway signal engineers of Saxby & Farmer. The design combined three semaphore arms with red and green gas lamps for night-time use, on a pillar, operated by a police constable. The gas lantern was manually turned by a traffic police officer, with a lever at its base so that the appropriate light faced traffic.

Although it was said to be successful at controlling traffic, its operational life was brief. It exploded on 2 January 1869, as a result of a leak in one of the gas lines underneath the sidewalk, injuring or killing the policeman who was operating it. With doubts about its safety, the concept was abandoned until electric signals became available.

An electric traffic light was developed in 1912 by Lester Wire, a policeman in Salt Lake City, Utah, who also used red-green lights. On 5 August 1914, the American Traffic Signal Company installed a traffic signal system on the corner of East 105th Street and Euclid Avenue in Cleveland, Ohio. It had two colors, red and green, and a buzzer, based on the design of James Hoge, to provide a warning for color changes. The design by James Hoge allowed police and fire stations to control the signals in case of emergency. The first four-way, three-color traffic light was created by police officer William Potts in Detroit, Michigan in 1920. Ashville, Ohio claims to be the home of the oldest working traffic light in the United States, used at an intersection of public roads from 1932 to 1982 when it was moved to a local museum. Many pictures of historical traffic lights appear at a Traffic Signal Trivia page.

Los Angeles installed its first automated traffic signals in October 1920 at five locations on Broadway. These early signals, manufactured by the Acme Traffic Signal Co., paired "Stop" and "Go" semaphore arms with small red and green lights. Bells played the role of today's amber or yellow lights, ringing when the flags changed—a process that took five seconds. By 1923 the city had installed 31 Acme traffic control devices. The Acme semaphore traffic lights were often used in Warner Bros. Looney Tunes and Merrie Melodies cartoons for comedic effect due to their loud bell.

The first interconnected traffic signal system was installed in Salt Lake City in 1917, with six connected intersections controlled simultaneously from a manual switch. Automatic control of interconnected traffic lights was introduced March 1922 in Houston, Texas.

After witnessing an accident between an automobile and a horse-drawn carriage, African American inventor, Garrett Morgan, became one of the first to file a U.S. patent for a traffic signal. Patent No. 1,475,024 was granted on November 20, 1923 for Morgan's three-position traffic signal - the ancestor of today's yellow warning light position commonly seen around the world today.

The first traffic lights in Britain were deployed in Piccadilly Circus in 1926. Outside of London, Wolverhampton was the first British town to introduce automated traffic lights in 1927 in Princes Square at the junction of Lichfield Street and Princess Street.

Melbourne was the first city in Australia to install traffic lights in 1928 on the intersection of Collins and Swanston Street.

Toronto, Ontario was the first city to computerize its entire traffic signal system, which it accomplished in 1963.

Countdown timers on traffic lights were introduced in the 1990s. Timers are useful for pedestrians, to plan whether there is enough time to cross the intersection before the end of the walk phase, and for drivers, to know the amount of time before the light turns green. In the United States, timers for vehicle traffic are prohibited, but pedestrian timers are now required on new or upgraded signals on wider roadways.

Types and Placement of Colors

Traffic lights can have several additional lights for filter turns or bus lanes. This one in Warrington, United Kingdom, also shows the red + yellow combination seen in a number of European countries, and a backing board with white border to increase the target value of the signal head. Improved visibility of the signal head is achieved during the night by using the retro-reflective white border.

Aspect is a term referring to the light that faces the drivers. A typical vertical traffic signal has three aspects, or lights, facing the oncoming traffic, red on top, yellow below, and green below that. Generally one aspect is illuminated at a time. In some cases, a fourth aspect, for a turn arrow for example, is below the three lights or aspects in more complicated road traffic intersections.

Single Aspects

The simplest traffic light comprises either a single or a pair of colored aspects that warns any user of the shared right of way of a possible conflict or danger.

- Flashing red: treated as a stop sign. This can also signal the road is closed. In France, flashing red mandates absolute stop, at the crossing of a railway line, an airport strip, or a swing bridge.
- Flashing yellow: caution, crossing or road hazard ahead.
- Flashing green: varies among jurisdiction. Flashing green can give permission to go straight as well as make a left turn in front of opposing traffic (which is stopped by a steady red light), can indicate the end of a green cycle before the light changes to a solid yellow, or (as in British Columbia, Canada) indicates the intersection is a pedestrian crosswalk. In Greece, flashing green lights cannot be found. Instead, dual flashing orange lights are used.

Dual Aspects

These have two lights, usually mounted vertically. They are often seen at railway crossings, fire stations, and intersections of streets. They flash yellow when cross traffic is not expected, and turn red to stop traffic when cross traffic occurs (e.g., the fire engines are about to exit the station). They are also sometimes used for ramp metering, where motorists enter a controlled-access highway during heavy traffic. Only one vehicle on the ramp proceeds when the signal shows green.

Three or More Aspects & Positioning of Aspects

The standard traffic signal is the red light above the green, with yellow between.

When the traffic signal with three aspects is arranged horizontally or sideways, the arrangement depends on the rule of the road. In right-lane countries, the sequence (from left to right) is red–yellow–green. In left-lane countries, the sequence is green–yellow–red.

Other signals are sometimes added for more control, such as for public transportation and right or left turns allowed only when the green arrow is illuminated or specifically prohibited if the red arrow is illuminated.

Traffic Signal Cycles

Generally, at least one direction of traffic at an intersection has the green light (green aspect) at any moment in the cycle. In some jurisdictions, for a brief time, all signals at an intersection show red at the same time, to clear any traffic in the intersection. The delay can depend on traffic, road conditions, the physical layout of the intersection, and legal requirements. Thus modern signals are built to allow the "all red" in an intersection, even if the feature is not used.

Some signals have no "all red" phase: the light turns green for cross traffic the instant the other light turns red.

Variations

In the Canadian province of Quebec and the Maritime provinces, lights are often arranged horizontally or sideways, but each aspect is a different shape: red is a square (larger than the normal circle) and usually in pairs at either end of the fixture, yellow is a diamond, and green is a circle. In many southern and southwestern U.S. states, most traffic signals are similarly horizontal in order to ease wind resistance during storms and hurricanes. The Canadian province of Alberta also uses this horizontal arrangement.

In Britain, normal traffic lights follow this sequence:

- Red – stop
- Red and yellow – stop, indicating it will turn green
- Green – proceed with caution, but only if the way is clear
- Yellow – stop if possible to do so without braking so sharply such as may cause an accident

Use of traffic signals in waterways, on railroads for rail traffic

The three-aspect standard is also used at locks on the Upper Mississippi River. Red means that another vessel is passing through. Yellow means that the lock chamber is being emptied or filled to match the level of the approaching vessel. After the gate opens, green means that the vessel may enter.

Railroad signals, for stopping trains in their own right of way, use the opposite positioning of the colors, the two types cannot be confused. That is, green on top and red below is the standard placement of the signal colors on railroad tracks.

Pedestrian and Cyclist Crossing Lights

Pedestrian crossing control panel in Dubai

Sounds

In some jurisdictions such as Australia and Ireland pedestrian lights are associated with a sound device, for the benefit of blind and visually impaired pedestrians. These make a slow beeping sound when the pedestrian lights are red and a continuous buzzing sound when the lights are green. In the Australian States of Queensland, New South Wales, Victoria and Western Australia, the sound is produced in the same unit as the push buttons. This system of assistive technology is also widely used at busy intersections in Canadian cities.

The United Kingdom, the Puffin crossings and their predecessor, the Pelican crossing, will make a fast beeping sound to indicate that it is safe to cross the road. The beeping sound is disabled during the night time so as not to disturb any nearby residents.

In some states in the United States, at some busy intersections, buttons will make a beeping sound for blind people. When the light changes, a speaker built into the button will play a recording to notify blind people that it is safe to cross. When the signal flashes red, the recording will start to count down with the countdown timer.

Australia and New Zealand

In Australia and New Zealand, the light sequence is:

- Green man: safe to cross the intersection
- Flashing red man: continue to cross if already in the intersection, but do not start to cross
- Red man: do not cross

Some traffic lights in Melbourne, Adelaide, Perth and Brisbane have countdown timers for pedestrian crossing lights, usually a countdown from 30 when the red flashing man appeared. Countdown timers have been installed on some pedestrian crossing lights in Auckland counting down from 15 seconds.

Some traffic-light controlled junctions have a light sequence that stops all vehicular traffic at the junction at the same time, and gives pedestrians exclusive access to the intersection so that they can cross in any direction (including diagonally). This is prominently featured in the Sydney CBD outside town hall. This is known as a “pedestrian scramble” or Barnes Dance in some places. In New Zealand such pedestrian crossings can be found in the central business district (CBD) of Auckland, Wellington and Christchurch.

The Barnes Dance is named after an American traffic engineer, Henry A. Barnes. Barnes did not claim to have invented the system but was a strong advocate of it, having observed the difficulties his daughter faced crossing the road to get to school.

European Standard

The European approach to a signalized crossing is to use dual or, more rarely, a triple aspect with a blackened out lens of a pictogram pedestrian. For cyclists, the same approach is used, with the lens blackened out for a bicycle frame. It is not uncommon to see lenses with both symbols on them. Most European countries use orange instead of yellow for the middle light.

Traffic light animation (pedestrians/cyclers and traffic). Ljubljana, Slovenia

Bicycle traffic lights in Vienna

The light sequence is:

- Green: safe to cross.
- Yellow or orange: continue to cross only if unable to stop safely.
- Flashing yellow or orange: cross with caution (often used when lights are out of order or shut down).
- Red: do not cross.

In Germany, the Czech Republic and some other Central European countries, a combination of red and orange lights is used just before the switch back to green. It allows drivers to stop their engines during the red light. The light sequence is as follows:

- Green: safe to cross.
- Orange: continue to cross only if unable to stop safely.
- Flashing orange: cross with caution, obey signage (used when lights are out of order or shut down).
- Red: do not cross.
- Red and orange: do not cross, prepare for green.

The light is blackened out with a pedestrian pictogram.

Ampelmännchen pedestrian traffic signals have come to be seen as a nostalgic sign for the former German Democratic Republic.

Variations

2007 in Prague Roman Tic (pseudonym) of the arts collective Ztohoven has changed 50 pedestrian traffic lights to different own designs of men and women, showing different actions and emotions, and subsequently has been fined.

In a stride for equality, some of individual men on the traffic lights have been changed to women or couples (both heterosexual and homosexual) Ampelpärchen in 2015 in Vienna, Salzburg and Linz (redone in December 2015).

British Standard

In the United Kingdom, Ireland, British Crown dependencies and dependent territories, and former possessions like Hong Kong two or more of the following signals are displayed to pedestrians -

- A still image of a green walking person: cross the road
- Flashing green walking person: continue to cross if already on the crossing but do not start to cross.
- Some signals use an amber numerical countdown display, instead of the flashing green man, to show how long people already crossing have remaining before the red standing man shows, and traffic is given a green signal
- Red standing man: do not cross/do not start to cross

In the United Kingdom there is no direct offense committed if a pedestrian fails to obey crossing signals and many lights commonly only use two still images - a green walking person and a red standing man, this being the general case where the crossing is at a road junction and the pedestrian signals are in combination with those controlling vehicular traffic. Flashing amber lights and images at pedestrian crossings are used where the vehicular traffic lights perform the sole function of stopping road traffic to allow pedestrians to cross a road.

The same system is used also in Switzerland and Macau.

China Standard

- Green: safe to cross.
- Red: do not cross.
- Yellow (solid, after green, before red): continue to cross only if unable to stop safely.
- Flashing yellow: cross with caution (often used in low-traffic crossing or after midnight).

North American Standard

In the United States, the most common aspect is the written 'walk' or 'don't walk.' In Canada, the white walking person is almost always used. Increasingly for retrofits of dual aspects and newer installations, the lower aspect formerly used for the "walk" signal (a walking person) is being replaced with a timer countdown. The raised hand pictogram first appeared in the city of Westmount, Montreal, and was invented by Alexander Heron, having observed a policeman's gloved hand controlling a crossing.

The light sequence is:

- Green, blue, or white walking person or "walk": cross with caution (pedestrians have the right of way; motorists turning left or right must yield to pedestrians).
- Flashing red or orange stop hand or "don't walk": do not start crossing, but continue if already in the middle of the intersection.
- Red or orange stop hand or "don't walk": do not enter the intersection

The U.S. state of Massachusetts allows an unusual indication variation for pedestrian movement. At signalized intersections without separate pedestrian signal heads, the traffic signals may be programmed to turn red in all directions, followed by a steady display of yellow lights simultaneously with the red indications. During this red-plus-yellow indication, the intersection is closed to vehicular traffic and pedestrians may cross, usually in whatever direction they choose (this is known as a "Barnes dance").

Lights for Public Transport

Four-lamp public transport traffic light in Moscow, Russia, showing "stop" signal.

A tram signal in Genèva, Switzerland.

TTC Spadina streetcar signals (smaller black signals)

Signals for buses and trams in Karlsruhe, Germany

Swedish traffic light (left) for use by public transport vehicles only. All signals use white lighting and special symbols ("S","–" and an arrow) to distinguish them from regular signals. The small light at the top tells the driver when the vehicle's transponder signal is received by the traffic light

Traffic lights for public transport often use signals that are distinct from those for private traffic. They can be letters, arrows or bars of white or colored light.

In Portland, Oregon, the tram signals feature an orange horizontal bar and a white vertical bar. Some systems use the letter B for buses, and T for trams.

In some European countries and Russia, dedicated traffic signals for public transport (tram, as well any that is using a dedicated lane) have four white lights that form the letter T. If the three top lamps are lit, this means "stop". If the bottom lamp and some lamps on the top row are lit, this means permission to go in a direction shown. In case of a tram signal, if there are no tram junctions or turns on an intersection, a simpler system of one yellow signal in the form of letter T is used instead; the tram must proceed only when the signal is lit.

In North European countries, the tram signals feature white lights of different forms: "S" for "stop", "—" for "caution" and arrows to permit passage in a given direction.

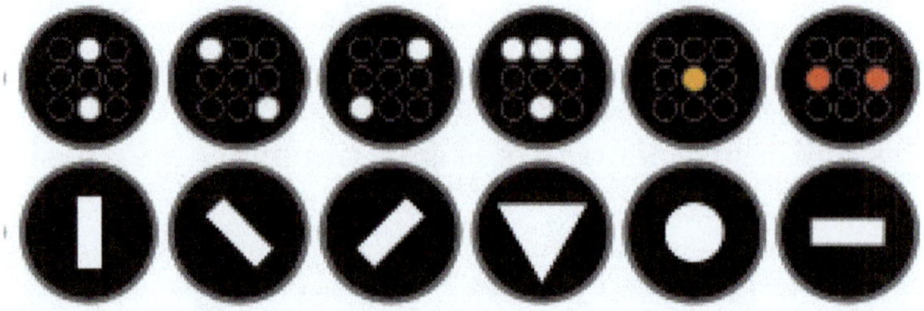

The Netherlands use a distinctive "negenoog" (nine-eyed) design shown on the top row of the diagram; bottom row signals are used in Belgium, Luxembourg, France and Germany. The signals mean (from left to right): "go straight ahead", "go left", "go right", "go in any direction" (like the "green" of a normal traffic light), "stop, unless the emergency brake is needed" (equal to "yellow"), and "stop" (equal to "red").

The METRO light rail system in Minneapolis, Minnesota as well as the RTA Streetcar System in New Orleans use a simplified variant of the Belgian/French system in the respective city's central business district where only the "go" and "stop" configurations are used. A third signal equal to yellow is accomplished by flashing the "go" signal.

In Japan, tram signals are under the regular vehicle signal; however the color of the signal intended for trams is orange.

In Hong Kong Island, Hong Kong, a yellow T-signal is used for trams, in place of the green signal. Addition to that, at any tramway junction, another set of signals is available to indicate the direction of the tracks.

In Australia a white "B" or "T" sometimes replaces the green light indicating that buses or trams (respectively) have right of way.

Preemption and Priority

Some regions have signals that are interruptible, giving priority to special traffic usually emergency vehicles such as fire apparatus, ambulances, and police squad cars. Most of the systems operate with small transmitters that send radio waves, infrared signals, or strobe light signals that are received by a sensor on or near the traffic lights. Some systems use audio detection, where a certain type of siren must be used and detected by a receiver on the traffic light structure.

Upon activation the normal traffic light cycle is suspended and replaced by the "preemption sequence": the traffic lights to all approaches to the intersection are switched to "red" with the exception of the light for the vehicle that has triggered the preemption sequence. Sometimes, an additional signal light is placed nearby to indicate to the preempting vehicle that the preempting sequence has been activated and to warn other motorists of the approach of an emergency vehicle.

The normal traffic light cycle resumes after the sensor has been passed by the vehicle that triggered the preemption.

In lieu of preemptive mechanisms, in most jurisdictions, emergency vehicles are not required to respect traffic lights. However, emergency vehicles must slow down, proceed cautiously and activate their emergency lights to alert oncoming drivers to the preemption when crossing an intersection against the light.

Unlike preemption, which immediately interrupts a signal's normal operation to serve the preempting vehicle and is usually reserved for emergency use, "priority" is a set of strategies intended to reduce delay for specific vehicles, especially mass transit vehicles such as buses. A variety of strategies exist to give priority to transit but they all generally work by detecting approaching transit vehicles and making small adjustments to the signal timing. These adjustments are designed to either decrease the likelihood that the transit vehicle will arrive during a red interval or decrease the length of the red interval for those vehicles that are stopped. Priority does not guarantee that transit vehicles always get a green light the instant they arrive like preemption does.

Turning Signals and Rules

A traffic light in Westbrook, Maine, on State Route 25. Notice the red arrow to the left of the two green straight lights.

In some instances, traffic may turn left (in left-driving jurisdictions) or right (in right-driving jurisdictions) after stopping at a red light, providing they give way to the pedestrians and other vehicles. In some places that generally disallow this, a sign next to the traffic light indicates that it is allowed at a particular intersection. Conversely, jurisdictions that generally allow this might forbid it at a particular intersection with a "no turn on red" sign, or put a green arrow to indicate specifically when a turn is allowed without having to yield to pedestrians (this is usually when traffic from the perpendicular street is making a turn onto one's street and thus no pedestrians are allowed in the intersection anyway). Some jurisdictions allow turning on red in the opposite direction (left in right-driving countries; right in left-driving countries) from a one-way road onto another one-way road; some of these even allow these turns from a two-way road onto a one-way road. Also differing is whether a red arrow prohibits turns; some jurisdictions require a "no turn on red" sign in these cases. A study in the State of Illinois (a right-driving jurisdiction) concluded that allowing drivers to proceed straight on red after stopping, at specially posted T-intersections where the intersecting road went left only, was dangerous. Proceeding straight on red at T-intersections where the intersecting road went left only used to be legal in Mainland China, with right-hand traffic provided that such movement would not interfere with other traffic, but when the Road Traffic Safety Law of the People's Republic of China took effect on 1 May 2004, such movement was outlawed. In some other countries, the permission is indicated by a flashing yellow arrow (cars do not have to stop but must give way to other cars and pedestrians).

Another distinction is between intersections that have dedicated signals for turning across the flow of opposing traffic and those that do not. Such signals are called dedicated left-turn lights in the United States and Canada (since opposing traffic is on the left). With dedicated left turn signals, a left-pointing arrow turns green when traffic may turn left without opposing traffic and pedestrian conflict, and turns red or disappears otherwise. Such a signal is referred to as a "protected" signal if it has its own red phase; a "permissive" signal does not have such a feature. Three standard versions of the permissive signal exist: One version is a horizontal bar with five lights – the green and yellow arrows are located between the standard green and yellow lights. A vertical five-light bar holds the arrows underneath the standard green light (in this arrangement, the yellow arrow is sometimes omitted, leaving only the green arrow below the solid green light, or possibly an LED based device capable of showing both green and yellow arrows within a single lamp housing). Some newer LED turn arrows seen in parts of Canada are capable of multicolored animation. Such lights will often display a flashing and animated green or yellow arrow when the dedicated turn is allowed, but then transform into a red arrow on a white background with a red line through it, emphasising that the turn is no longer allowed. These lights will also often have the words "no turn" displayed, or an explanatory reason why the turn is not allowed, such as "train" in the case of a rail or light rail crossing. A third type is known as a "doghouse" or "cluster head" – a vertical column with the two normal lights is on the right side of the signal, a vertical column with the two arrows is located on the left, and the normal red signal is in the middle above the two columns. Cluster signals in Australia and New Zealand use six signals, the sixth being a red arrow that can operate separately from the standard red light. In a fourth type, sometimes seen at intersections in Ontario and Quebec, Canada, there is no dedicated left-turn lamp per se. Instead, the normal green lamp flashes rapidly, indicating permission to go straight as well as make a left turn in front of opposing traffic, which is being held by a steady red lamp. (This "advance green," or flashing green can be somewhat startling and confusing to drivers not familiar with this system. This also can cause confusion amongst visitors to British Columbia, where a flashing green signal denotes a pedestrian controlled crosswalk.) Another interesting practice seen at least in Ontario is that cars wishing to turn left that arrived after the left turn signal ended can do so during the yellow phase, as long as there is enough time to make a safe turn.

A flashing yellow arrow, which allows drivers to make left turns after giving way to oncoming traffic, is becoming more widespread in the United States, particularly in Oregon, North Carolina, Michigan (replacing their trademark "red-ball" flashing left-turn lights), and Las Vegas, Nevada. In the normal sequence, a protected green left-turn arrow will first change to a solid yellow arrow to indicate the end of the protected phase, then to a flashing yellow arrow, which remains flashing until the standard green light changes to yellow and red. These generally take the form of four signal sections (green, yellow, yellow arrow, red). On some newer signals, notably in the city of Bend, Oregon, the green and flashing yellow arrows emanate from the same light section through the use of a dual-color LED array, while the solid yellow arrow is mounted above it. In Las Vegas, the arrow turns flashing yellow from a red light, before turning red again.

Generally, a dedicated left-turn signal is illuminated at the beginning of the green phase of the green-yellow-red-green cycle. This is called a leading turn. This allows left-turn traffic, which often consists of just a few cars, to vacate the intersection quickly before giving priority to vehicles traveling straight. This increases the throughput of left-turn traffic while reducing the number of drivers, perhaps frustrated by long waits in heavy traffic for opposing traffic to clear, attempting to make an illegal left turn on red. A dedicated left-turn signal that appears at the end of the green phase is called a lagging turn. If there is no left-turn signal, the law requires one to yield to oncoming traffic and turn when the intersection is clear and it is safe to do so. Nevertheless, it is increasingly common in the U.S. to see drivers who do not yield in the absence of a dedicated signal, cutting off traffic that has the right-of-way and is starting to head across the intersection. In the U.S., many older inner-city and rural areas do not have dedicated left-turn lights, while most newer suburban areas have them. Such lights tend to decrease the overall efficiency of the intersection as it becomes congested, although it makes intersections safer by reducing the risk of head-on collisions and may even speed up through traffic, but if a significant amount of traffic is turning, a dedicated turn signal helps eliminate congestion.

Some intersections with permissive turn signals occasionally have what is known as "yellow trap", "lag-trap", or "left turn trap" (in right-driving countries). This refers to situations when left-turning drivers are trapped in the intersection with a red light, while opposing traffic still has a green. In British Columbia, the law addresses this problem by giving a left-turning driver already in an intersection the right-of-way to make the turn once the intersection is otherwise clear, regardless of the traffic light state.

For example, an intersection has dedicated left-turn signals for traffic traveling north. The southbound traffic gets a red light so northbound traffic can make a left turn, but the straight-through northbound traffic continues to get a green light. A southbound driver who had entered the intersection earlier will now be in a predicament, since they have no idea whether traffic continuing straight for both directions is becoming red, or just their direction. The driver will now have to check the traffic light behind them, which is often impossible from the viewing angle of a driver's seat. This can also happen when emergency vehicles or railroads preempt normal signal operation. In the United States, signs reading "Oncoming traffic has extended green" or "Oncoming traffic may have extended green" must be posted at intersections where the "yellow trap" condition exists.

Although motorcycles and scooters in most jurisdictions follow the same traffic signal rules for left turns as do cars and trucks, some places, such as Taiwan, have different rules. In these areas, it is not permitted for such small and often hard-to-see vehicles to turn left in front of oncoming traffic on certain high-volume roads when there is no dedicated left-turn signal. Instead, in order to make a left turn, the rider moves to the right side of the road, travels through the first half of the intersection on green, then slows down and stops directly in front of the line of cars on the driver's right waiting to travel across the intersection, which are of course being held by a red light. There is often a white box painted on the road in this location to indicate where the riders should group. The rider turns the bike 90 degrees to the left from the original direction of travel and proceeds along with the line of cars when the red light turns green, completing the left turn. This procedure improves safety because the rider never has to cross oncoming traffic, which is particularly important given the much greater likelihood of injury when a cycle is hit by a car or truck. This system (called a "hook turn") is also used at many intersections in the CBD of Melbourne, Australia, where either or both streets carry tramways. This is done so right-turning vehicles (Australia drives on the left) do not block the passage of trams. The system is being extended to the suburbs.

Lane Control

Lane-control lights are a specific type of traffic light used to manage traffic on a multi-way road, highway or tollway. Typically, these lights allow or forbid traffic to use one or more of the available lanes by the use of green lights or arrows (to permit) or by red lights or crosses (to prohibit). In the US, lane-control lights are often used to control and/or direct the flow of traffic through toll plazas and highway tunnels, such as during unusually-heavy traffic flow when more lanes may be required in one direction than in the other direction, or during a hurricane evacuation, when the lane signals for most or all lanes will show green for one direction to assist in more rapid traffic flow from the evacuation site. Lane-control lights are also used at highway weigh stations to direct tractor-trailers and other heavy or oversized vehicles into the proper lanes for weighing, inspection or exit.

In the US, most notably the Southeastern, there often is a "continuous-flow" lane. This lane is protected by a single, constant-green arrow pointing down at the lane(s) permitting the continuous flow of traffic, without regard to the condition of signals for other lanes or cross streets. Continuous lanes are restricted in that vehicles turning from a side street may not cross over the double white line to enter the continuous lane, and no lane changes are permitted to the continuous lane from an adjacent lane or from the continuous lane to an adjacent lane, until the double white line has been passed. Some continuous lanes are protected by a raised curb located between the continuous lane and a normal traffic lane, with white and/or yellow reflective paint or tape, prohibiting turning or adjacent traffic from entering the lane. Continuous-flow traffic lanes are found only at "T" intersections where there is no side street or driveway entrance on the right side of the main thoroughfare; additionally, no pedestrians are permitted to cross the main thoroughfare at intersections with a continuous-flow lane, although crossing at the side street may be permitted. Intersections with continuous-flow lanes will be posted with a white regulatory sign approximately 500 ft (150 m) before the intersection with the phrase, "right lane continuous traffic," or other, similar, wording. If the arrow is extinguished for any reason, whether by malfunction or design, traffic through the continuous lane will revert to the normal traffic pattern for adjacent lanes, except that turning or moving into or out of the restricted lane is still prohibited.

Special Provisions

Traffic light failure in most jurisdictions in both drive-on-the-left Australia and some states of the mainly drive-on-the-right Europe must be handled by drivers as a priority-to-the-right intersection, or an all-way stop elsewhere, pending the arrival of a police officer to direct traffic. Some jurisdictions, however, have additional right-of-way signs mounted above, below or next to the traffic lights; these take effect when the lights are no longer active. In the UK and parts of North America, drivers simply treat the junction as being uncontrolled when traffic lights fail, giving way as appropriate, unless a police officer is present. In much of the United States failed traffic signals must be treated as all-way stop intersections.

In Italy as well as some jurisdictions in the US, traffic lights inactive at nighttime emit a yellow-colored flashing signal in directions owing priority while the intersecting street emit a flashing red light, requiring drivers to stop before proceeding. In Germany the priority directions will not be illuminated while the intersecting streets will be shown a flashing yellow signal.

Dummy Lights

Historic dummy light in Canajoharie, New York, United States

In an era when intersections were often controlled by a single traffic signal head, many signals were installed on pedestals in the centers of intersections. Often referred to as "dummy lights," these installations often replaced beacons or "mushrooms" that denoted the centers of intersections and separated opposing traffic, with the infrastructure used for the beacons and mushrooms serving the new "stop and go" type signals.

There are a handful of operational dummy lights still in service. Three are located in New York State: Beacon, Canajoharie and Croton-on-Hudson.

After a dummy light was knocked down by a truck in 2010, the city of Coleman, Texas decided to preserve and refurbish its last two pedestal mounted dummy lights as part of its historic district preservation efforts.

In 2011, the Arkansas State Historic Preservation Office nominated the state's last remaining pedestal mounted signal, located in Smackover, to be listed on the National Register of Historic Places.

The oldest working 'dummy' style beacon in Rhode Island was located in the Historic Mill village of Albion in the Blackstone River Valley (Town of Lincoln RI). It was erected in 1932 above where the old village well stood in the center square. In April 2015 it was destroyed by a motorist. The beacon was a local historical site and the logo for a local Boy Scout Troop, Troop 711 Albion.

Increases in traffic flows have prompted calls for these types of traffic lights to be removed due to safety concerns, but their historic value has kept these landmarks at their original locations. To serve historic district applications, Teeco Safety Systems of Shreveport, Louisiana, still manufactures replacement fixed 4-way traffic signals for pedestal and overhead span wire installations.

Technology

Optics and Lighting

An example of a LED traffic light in Australia

Traditionally, incandescent and halogen bulbs were used. Because of the low efficiency of light output and a single point of failure (filament burnout) municipalities are increasingly retrofitting traffic signals with LED arrays that consume less power, have increased light output, last significantly longer, and in the event of an individual LED failure, still operate albeit with a reduced light output. With the use of optics, the light pattern of an LED array can be comparable to the pattern of an incandescent or halogen bulb.

The low energy consumption of LED lights can pose a driving risk in some areas during winter. Unlike incandescent and halogen bulbs, which generally get hot enough to melt away any snow that may settle on individual lights, LED displays – using only a fraction of the energy – remain too cool for this to happen. As a response to the safety concerns, a heating element on the lens was developed.

Programmable Visibility Signals

Traffic signals installed in Shelton, Washington. Standing off-axis from the intended viewing area, these signals appear to be "off" or invisible to adjacent lanes of traffic during the daytime. Only a faint glow can be seen when viewed at night

Signals such as the 3M High Visibility Signal and McCain Programmable Visibility signal utilize light-diffusing optics and a Fresnel lens to create the signal indication. The light from a 150 W PAR46 sealed-beam lamp in these "programmable visibility" signals passes through a set of two glass lenses at the back of the signal. The first lens, a frosted glass diffusing lens, diffuses the light into a uniform ball of light around five inches in diameter. The light then passes through a nearly identical lens known as an optical limiter (3M's definition of the lens itself), also known as a "programming lens", also five inches in diameter.

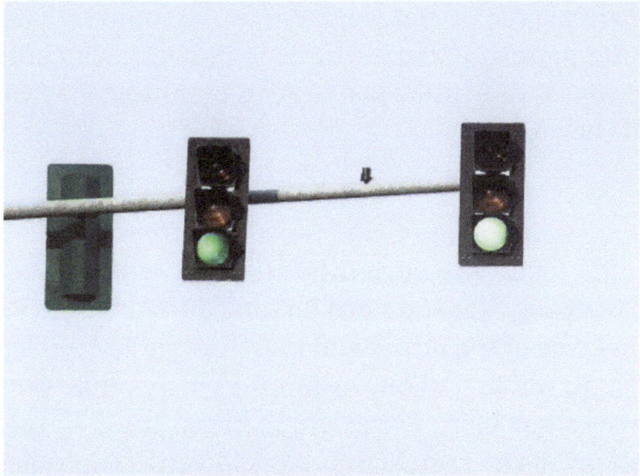

The same two signals above, taken in the signal's intended viewing area (a single lane of northbound traffic). Special light-diffusing optics and a colored Fresnel lens create the indication

Using a special aluminum foil-based adhesive tape, these signals are "masked" or programmed by the programming lens so that only certain lanes of traffic will view the indication. At the front of these programmable visibility signals is a 12" Fresnel lens, each lens tinted to meet United States Institute of Transportation Engineers (ITE) chromaticity and luminance standards. The Fresnel lens collimates the light output created by the lamp, and creates a uniform display of light for the lane in which it is intended. These signals were first developed by the 3M Company in the late 1960s, and were popular in the late 1970s as traffic density increased.

In addition to being positioned and mounted for desired visibility for their respective traffic, some traffic lights are also aimed, louvered, or shaded to minimize misinterpretation from other lanes. For example, a Fresnel lens on an adjacent through-lane signal may be aimed to prevent left-turning traffic from anticipating its own green arrow.

Today, McCain Traffic Systems is the only U.S.-based manufacturer producing optically programmable traffic signals similar to the 3M model. Intelight Inc. manufactures a programmable traffic signal that uses a software-controlled LED array and electronics to steer the light beam toward the desired approach. The signal is programmed unlike the 3M and McCain models. It requires a connection to a laptop or smartphone with the manufacturer software installed. Connections can be made directly with a direct-serial interface kit, or wirelessly with a radio kit over WIFI to the signal.

In addition to aiming, Fresnel lenses, and louvers, visors and back panels are also useful in areas where sunlight would diminish the contrast and visibility of a signal face.

Typical applications for these signals were skewed intersections, specific multi-lane control, left-turn pocket signals or other areas where complex traffic situations existed.

Conventional Lighting Systems

Conventional traffic signal lighting, still common in some areas, utilizes a standard light bulb. Typically, a 67, 69, or 115 watt medium-base (household lamp in the US) light bulb provides the illumination. Light then bounces off a mirrored glass or polished aluminium reflector bowl, and out through a polycarbonate plastic or glass signal lens. In some signals, these lenses were cut to include a specific refracting pattern. Crouse-Hinds is one notable company for this practice. In the 1930s throughout the 1950s, they utilized a beaded prismatic lens with a "smiley" pattern embossed into the bottom of each lens.

Light Design

In the United States, traffic lights are currently designed with lights approximately 12 inches (300 mm) in diameter. Previously the standard had been 8 inches (200 mm), however those are slowly being phased out in favor of the larger and more visible 12 inch lights. Variations used have also included a hybrid design, which had one or more 12 inch lights along with one or more lights of 8 inches (200 mm) on the same light. For example, these "12-8-8" (along with 8-8-8) lights are standard in most jurisdictions in Ontario, Manitoba, and British Columbia (that, is, the red light is 12 and others 8, making the red more prominent).

In the United Kingdom, 12 inch lights were implemented only with Mellor Design Signal heads designed by David Mellor. These were designed for symbolic optics to compensate for the light loss caused by the symbol. However, following a study sponsored by the UK Highways Agency and completed by Aston University, Birmingham, UK, an enhanced optical design was introduced in the mid 1990s. Criticism of sunlight *washout* (cannot see the illuminated signal due to sunlight falling on it), and *sun-phantom* (signal appearing to be illuminated even when not due to sunlight reflecting from the parabolic mirror at low sun angles), led to the design of a signal that used lenslets to focus light from a traditional incandescent bulb through apertures in a matt black front mask. This cured both problems in an easily manufactured solution. This design proved very successful and was taken in to production by a number of traffic signal manufacturers through the engineering designs of Dr Mark Aston, working firstly at the SIRA Ltd in Kent, and latterly as an independent optical designer. The manufacturers took a licence for the generic design from the Highways Agency, with Dr Aston engineering a unique solution for each manufacturer. Producing both bulb and LED versions of the signal aspects, these signals are still the most common type of traffic light on UK roads. With the invention of anti-phantom, highly visible Aston lenses, lights of 8 inches (200 mm) could be designed to give the same output as plain lenses, so a larger surface area was unnecessary. Consequently, lights of 12 inches (300 mm) are no longer approved for use in the UK and all lights installed on new installations have to be 200 mm (8 in) in accordance with TSRGD (Traffic Signs Regulations and General Directions). Exemptions are made for temporary or replacement signals.

Technological Advancements

With technologies in developed countries continuing to advance, there is now an increasing move- to develop and implement smart traffic lights on the roads. These are systems that adapt to information that is received from a central computer about the position, speed and direction of vehicles. They try to communicate with cars to alert drivers of impending light changes and reduce motorists' waiting time considerably. Trials are currently being conducted for the implementation of these advanced traffic lights but there are still many hurdles to widespread use that need to be addressed; one of which is the fact that few cars yet have the required systems to communicate with these lights.

Control and Coordination

The normal function of traffic lights requires sophisticated control and coordination to ensure that traffic moves as smoothly and safely as possible and that pedestrians are protected when they cross the roads.

Mounting

There are significant differences from place to place in how traffic lights are mounted or positioned so that they are visible to drivers. Depending upon the location, traffic lights may be mounted on poles situated on street corners, hung from horizontal poles or wires strung over the roadway, or installed within large horizontal gantries that extend out from the corner and over the right-of-way. In the last case, such poles or gantries often have a lit sign with the name of the cross-street.

- Mounted traffic lights in the United States

An example of a wire-mounted traffic light in Fort Worth, Texas.

A typical traffic light on Ninth Avenue in New York City. Note that the traffic light gantry is of a guy-wire masted style, which is unique to New York City.

A vertically mounted traffic light in Boston with pole-mounted traffic light in the foreground.

In some locations, lights are mounted with their multiple faces arranged horizontally, often with supplemental vertical signals on the side, while others locations use vertical signals almost exclusively. Horizontal signals have consistent orientation, like their vertical counterparts. Often, supplemental curb pedestal mounts, intended to support a signal for a different approach road, are used when primary signals are partially obscured due to structures such as overpasses, approaches around a building that obscure the primary signal mountings, and unusual approach geometry. In Florida, horizontal signals mounted on poles, known as "mast arms", are in wide use due to their lower wind profile, important for minimizing hurricane damage. In areas where wind-load is not as much of a concern as ice-load, such as Illinois or Minnesota, the lights are mounted vertically to reduce the accumulation of ice or snow over the surface of the signal heads. In a few countries such as Japan, South Korea, Mexico and a few jurisdictions in Canada and the US such as Texas, New Mexico, Florida, Wisconsin, Nebraska, Quebec, and Alberta, most traffic signals are mounted horizontally.

Traffic signals in most areas of Europe are located at the stop line on same side of the intersection as the approaching traffic (there being both right- and left-hand traffic) and are often mounted overhead as well as on side of the road. At particularly busy junctions for freight, higher lights may be mounted specifically for trucks. The stop line alignment is done to prevent vehicles blocking any crosswalk and allow for better pedestrian traffic flow. There may also be a special area a few meters in advance of the stop line where cyclists may legally wait but not motor vehicles; this advanced stop line is often painted with a different road surface with greater friction and a high color, both for the benefit of cyclists and for other vehicles. The traffic lights are mounted so that cyclists can still see them.

In Spain, the mounted traffic lights on the far side of the intersection is meant for the traffic that exits the intersection in that particular direction. This is often done due to the pedestrian crossings, so that traffic has to wait if they get a red light. These intersections also come with a stop line in the exit area of the intersection.

In North America, there is often a pole-mounted signal on the same side of the intersection, but additional pole-mounted and overhead signals are usually mounted on the far side of the intersection for better visibility. Most traffic lights are mounted that way in the Western United States and Canada. In Ontario, traffic lights are almost always mounted on the far side of the intersection with poles.

In some areas of the United States, signals facing in up to four directions are hung directly over the intersection on a wire strung diagonally over the intersection (once common in Michigan), or the signal faces traffic in one direction, still hung by wires (but the wire is strung horizontally between two adjacent corners of the intersection). This is common in the Southern and Eastern United States.

In other countries like Australia, New Zealand, UK, traffic lights are mounted at the stop line before the intersection and also after the intersection. Some busy intersections have an overhead traffic light for heavy vehicles and vehicles further away.

Implementation

According to transportation engineers, traffic lights can have both positive and negative effects on traffic safety and traffic flow. The separation of conflicting streams of traffic in time can reduce the chances of right-angle collisions. But also the frequency of rear-end crashes can be increased by the installation of traffic lights, and they can adversely affect the safety of bicycle and pedestrian traffic. They can increase the traffic capacity at intersections, but can also result in excessive traffic delay. Hans Monderman, the innovative Dutch traffic engineer, and pioneer of shared space schemes, was sceptical of their role, and is quoted as having said of them: “We only want traffic lights where they are useful and I haven’t found anywhere where they are useful yet.”

Small Vehicle Safety

In some instances the stoplight traffic detector will not change the light for small vehicles such as motorcycles, scooters and particularly bicycles, because the induction loop sensor fails to detect the small vehicle. A vehicle with sufficient mass of metal such as steel interacts with the sensor’s magnetic field causing the light to change at the appropriate time. Motorcycles and scooters have much less mass than cars, resulting in a failure to trigger the sensor, while bicycles may not even be constructed with metal. This situation most often occurs at the times of day when other traffic is sparse as well as when the small vehicle is coming from a direction that does not have a high volume of traffic. Not all sensor traffic lights have this problem. There are some that can pick up these vehicles without problems and most are capable of doing so. However, through road construction and other processes, they are not properly adjusted or calibrated for the small vehicles. Magnetic field sensors detect vehicles passing over them and not necessarily stationary vehicles.

Actuated Traffic Lights

Some states in the United States have passed laws which give motorcyclists an affirmative defense to proceed through a red light with caution after stopping when the device which causes the light to change from red to green does not activate. A partial listing of these state statutes include Wisconsin 346.37(1)(c)(4), Illinois Sec. 11-306(c)(3.5), Minnesota 169.06 Subd. 9, South Carolina S.C. Code 56-5-970 (5), North Carolina, Idaho, Arkansas, and Tennessee. Additional state laws regarding actuated traffic lights can be found at the American Motorcycle Association (AMA) website, including this quick reference guide.

Traffic Signal Warrants

Traffic signals have strengths and weaknesses that must be considered when deciding whether to install them. Signaled intersections can reduce delay for side road traffic and reduce the occurrence of collisions by turning traffic and cross traffic. But they may also cause delay for traffic on the main road, and often increase rear-end collisions by up to 50%. Since right-angled and turn-against-traffic collisions are more likely to result in injuries, this is often an acceptable trade-off.

United States MUTCD

Criteria have been developed to help ensure that new signals are installed only where they will do more good than harm. In the United States these criteria are called *warrants*, and are found in the Manual on Uniform Traffic Control Devices (MUTCD), the Federal regulation covering the use of signs, pavement markings, traffic signals, and similar devices.

In the US MUTCD, there are nine warrants for traffic signals:

- Eight-hour vehicular volume. Traffic volume must exceed prescribed minima for eight hours of an average day (usually a weekday).
- Four-hour vehicular volume. Traffic volume must exceed prescribed minima for four hours of an average day.
- Peak hour volume or delay. This is applied only in unusual cases, such as office parks, industrial complexes, and park and ride lots that attract or discharge large numbers of vehicles in a short time, and, for a minimum of one hour of an average day. The side road traffic suffers undue delays when entering or crossing the major street.
- Pedestrian volume. If the traffic volume on a major street is so heavy that pedestrians experience excessive delays in attempting to cross it.
- School crossing. If the traffic density at school crossing times exceeds one per minute which is considered to provide too few gaps in the traffic for children to safely cross the street.
- Coordinated signal system. For places where adjacent traffic control signals do not keep traffic grouped together efficiently.
- Crash experience. The volumes in the eight- and four-hour warrants may be reduced if five or more right-angle and cross traffic turn collisions have happened at the intersection in a twelve-month period.
- Roadway network. Installing a traffic control signal at some intersections might be justified to encourage concentration and organization of traffic flow on a roadway network.
- Intersection near a grade crossing. A traffic control signal is often justified at an intersection near a railroad crossing, in order to provide a preemption sequence to allow traffic queued up on the tracks an opportunity to clear the tracks before the train arrives.

An intersection is usually required to meet one or more of these warrants before a signal is installed. However, meeting one or more warrants does not require the installation of a traffic signal, it only suggests that they may be suitable. It could be that a roundabout would work better. There may be other unconsidered conditions that lead traffic engineers to conclude that a signal is undesirable. For example, it may be decided not to install a signal at an intersection if traffic stopped by it will back up and block another, more heavily trafficked intersection. Also, if a signal meets only the peak hour warrant, the advantages during that time may not outweigh the disadvantages during the rest of the day.

Between 1979 and 1988, the city of Philadelphia, Pennsylvania, removed signals at 199 intersections that did not meet any of the warrants. On average, the intersections had 24% fewer crashes after the unwarranted signals were removed. The traffic lights had been erected in the 1960s because of since-resolved protests over traffic. By 1992, over 800 traffic lights had been removed at 426 intersections, and the number of crashes at these intersections dropped by 60%.

Legal Implications

In virtually all jurisdictions in which they are used, it is an offense for motorists (and other road users) to disregard the instructions of traffic lights (or other traffic control devices). Exceptionally, it is not an offense for pedestrians to cross against a red light in the United Kingdom, where pedestrian lights officially give advice, rather than an instruction, although UK pedestrians do commit an offense if they cross a road against the signals of a police officer controlling traffic.

California attempt to discourage red light running by posting the minimum fine.

Perhaps the most obvious common traffic-light related offense is failing to stop for a red light (in some jurisdictions, running a yellow light may also incur a penalty). Enforcement of traffic lights varies from jurisdiction to jurisdiction; some places are extremely strict, while other locales are known for traffic lights being routinely ignored by motorists, with no serious attempts by law enforcement to alter the situation.

In some jurisdictions (such as Toronto, Washington, D.C., New York City, and California), there are ordinances or by-laws against "gridlocking". A motorist entering an intersection (even if on a green light) but unable to proceed and who gets stranded in the intersection (when traffic ahead fails to proceed), and who remains after the light turns red (thus blocking traffic from other directions) may be cited. The definition of the intersection area is that square where the two streets overlap marked by the inner lines of each crosswalk. (Occupying the space inside the crosswalk lines is itself a traffic infraction, but different from gridlocking). This gives the meaning to the anti-gridlock slogan "don't block the box". This is sometimes used as a justification for making a turn across the opposing travel lanes on a red light at a busy intersection, by pulling partway into the intersection at a green light waiting to perform the turn, and, if oncoming traffic is not abated before the light changes to red, proceeding to turn once the light has turned red and opposing traffic has stopped. This means that at busy junctions without a protected green arrow for turning traffic, one turns after the light turns red. This maneuver is commonly referred to as "occupying the intersection" or "being legally allowed to complete one's turn". In some jurisdictions, including most American states, a vehicle already in the intersection when the light turns red legally has the right of way, and vehicles who have green must yield to the vehicle in the intersection.

In Sackville, New Brunswick, it is customary for through traffic to voluntarily yield to the first oncoming left-turning vehicle to allow it to perform a "Pittsburgh left". This is similar to a hook turn performed in Melbourne, Australia, which is legal at signed intersections.

A driver comes to a stop on the crosswalk beyond the stop line as a result of attempting to avoid a red light violation

Enforcement of traffic lights is done in one of several ways:

- by police officers observing traffic, and issuing citations to motorists who violate the signal
- as a result of an accident investigation, if it is determined that one or more motorists ran the red light — even if the incident was not observed by a police officer
- with red light cameras

Red Light Running

Jurisdictions differ somewhat on how to deal with "red light running" — attempts by motorists to race to an intersection while facing a yellow light, in an attempt to beat the red. In some locales, as long as the light is yellow when the motorist enters the intersection, no offense has been committed; in others, if the light turns red at any time before a motorist clears the intersection, then an offense occurs. In New York City the yellow light is very short (only about a second) in order to discourage driving through. In Oregon and other places, a stricter standard applies—running a yellow light is an offense, unless the motorist is unable to stop safely. This standard has been criticized as ambiguous and difficult to enforce (red light cameras in Oregon are activated only if a motorist enters the intersection on a red). Red light cameras in New South Wales, Australia, are activated only if a motorist enters an intersection 0.3 seconds after the light has turned red.

As urban centers become more dense and vehicles and pedestrians come into closer contact with each other, the risk of crashes increases. With the rapid increase of vehicles for hire through smart phone apps and the competition from taxis and livery cabs, the urgency to complete as many rides in as short a period of time as possible has led to drivers pushing the limits on red lights. According to the IIHS Insurance Institute for Highway Safety, every year red light running causes hundreds of deaths, thousands of injuries and hundreds of millions of dollars in related costs.

In Spring 2015, Hunter College in New York City completed an observational study of red light running, the first of its kind. The conclusion, after monitoring 3,259 vehicles at 50 intersections over a period of days around the 5 boroughs, found that almost 10% of vehicles and 15% of taxis ran the red light, amounting to nearly 400,000 red lights run every single day. New York City Mayor Bill de Blasio launched a Vision Zero plan to reduce vehicular and pedestrian fatalities.

The New York City Police Department's 2014 TrafficStat report showed 33,577 red light tickets for 2013, which is 126% increase in the number of failure-to-yield summonses and red-light running violations.

The *Amber Gambler Twins* is one of many public interest films trying to change the public's behavior regarding running of yellow lights.

Red Light Cameras

In some countries, red light cameras are used. An automated camera is connected to the triggering mechanism for the corresponding traffic light, which is targeted to photograph a vehicle and driver crossing against the light. Either the driver or the vehicle's owner (depending on local laws) are fined for the violation. In some jurisdictions, including the United States and Italy, private companies have been contracted to operate traffic-related cameras and receive a portion of the resulting revenues. In some cases red light cameras have been abused by local governments, where vehicle operators have been fined as a result of traffic systems that have been improperly modified. Despite the fact that cameras can reduce the number of crashes, it has been proven that at these intersections drivers tended to react quicker to a yellow light change when stopping. The consequence of this change could be the slight decline in the intersection capacity.

Confirmation Lights

An intersection with blue confirmation lights in Newport News, Virginia

Another way police officers have begun to combat red light runners is with *Confirmation Lights*. These lights, usually blue or white, can be seen from anywhere in an intersection, making it so that officers don't have to have a line-of-sight with a red light to catch someone illegally entering an intersection. They are only on when the red light they are associated with is on. Some intersections will also have multiple confirmation lights for a single direction of travel if there are different signals for different directions. These lights are separate from the main ones, often protrude above or below the main traffic light, and are much smaller than a standard light to help avoid confusion.

In the Netherlands, many traffic signals that are red can be seen from the side via a small bulbous window, indicating to drivers (and police officers) whether the signal in the crossing direction is actually red or not, by simply leaking out some of the red light through the side of the traffic signal. This has gradually become less common as traditional incandescent signals are replaced by LED signals, while increasingly red light cameras are used to detect driving-through-red violations.

Light Timing Length

Traffic light with time in Tehran, Iran

The length of yellow lights can differ, for example in many places the length of a yellow light is usually four or five seconds, but elsewhere it may be as little as three, considerably reducing the time for reaction. It is typical for these times to vary according to the set speed limit, with longer times for higher limits. In the U.S. state of Georgia, a yellow light must be lit one second for every 10 miles per hour (16 km/h) of posted speed limit. For intersections with red light cameras, one extra second must be added. In Colorado Springs, Colorado, yellow lights at numerous intersections have been noted with durations of approximately two seconds. In the United States, there is a recommended federal safety minimum of three seconds for yellow lights.

The time from when a red light is displayed and when a cross street is given a green light is usually based on the physical size of the intersection. This intervening period is called the "all-red time". A typical all-red time is two seconds to allow cars to clear the intersection. In a wider intersection, such as a four-lane road or highway intersection, the all-red time may be as much as five seconds, allowing drivers who could not or would not stop at the yellow light enough time to clear the intersection without causing a collision. Two exceptions are in New Westminster, British Columbia, Canada, and Winnipeg, Manitoba, Canada, where there is no all-red time. The change is instantaneous, due to the nature of the older relay operated signals. It is also the case in Montreal, Quebec, Canada.

In Other Contexts

The symbolism of a traffic light (and the meanings of the three primary colors used in traffic lights) are frequently found in many other contexts. Since they are often used as single spots of color without the context of vertical position, they are typically not comprehensible to up to one in ten males who are color blind.

Traffic lights have also been used in computer software, such as the OS X user interface, and in pieces of artwork, particularly the Traffic Light tree in London, UK.

Racing

Automobile racing circuits can also use standard traffic signals to indicate to racing car drivers the status of racing. On an oval track, four sets may be used, two facing a straight-away and two facing the middle of the 180-degree turn between straight-away. Green would indicate racing is under way, while yellow would indicate to slow or while following a pace car; red would indicate to stop, probably for emergency reasons.

Scuderia Ferrari, a Formula One racing team, formerly used a traffic light system during their pit stops to signal to their drivers to when to leave the pits. The red light was on when the tires were being changed and fuel was being added, yellow was on when the tires were changed, and green was on when all work was completed. The system is (usually) completely automatic. However, the system was withdrawn after the 2008 Singapore Grand Prix, due to the fact that it heavily delayed Felipe Massa during the race, when he was in the lead. Usually, the system was automatic, but heavy traffic in the pit lane forced the team to operate it manually. A mechanic accidentally pressed the green light button when the fuel hose was still attached to the car, causing Massa to drive off, towing the fuel hose along. Additionally, Massa drove into the path of Adrian Sutil, earning him a penalty. He finally stopped at the end of the pit lane, forcing Ferrari's mechanics to sprint down the whole of the pit lane to remove the hose. As a result of this, and the penalty he also incurred, Massa finished 13th. Ferrari decided to use a traditional "lollipop" for the remainder of the 2008 season.

Another type of traffic light that is used in racing is the Christmas Tree, which is used in drag racing. The Christmas Tree has six lights: a blue staging light, three amber lights, a green light and a red light. The blue staging light is divided into two parts: Pre-stage and stage. Sometimes, there are two sets of bulbs on top of each other to represent them. Once a driver is staged at the starting line, then the starter will activate the light to commence racing, which can be done in two ways. If a Pro tree is used, then the three amber lights will flash at the same time. For the Sportsman tree, the amber light will flash from top to bottom. When the green light comes up, the race officially begins but if a driver crosses the line before that happens, then a red light will come up and that will be a foul.

As a Rating Mechanism

The colors red, yellow and green are often used as a simple-to-understand rating system for products and processes. It may be extended by analogy to provide a greater range of intermediate colors, with red and green at the extremes.

Variable-message Sign

A variable- (also changeable-, electronic-, or dynamic-) message sign, often abbreviated VMS, CMS, or DMS, and in the UK known as a matrix sign, is an electronic traffic sign often used on roadways to give travelers information about special events. Such signs warn of traffic congestion, accidents, incidents, roadwork zones, or speed limits on a specific highway segment. In urban areas, VMS are used within parking guidance and information systems to guide drivers to available car parking spaces. They may also ask vehicles to take alternative routes, limit travel speed, warn of duration and location of the incidents or just inform of the traffic conditions.

History

VMS's were deployed at least as early as the 1950s on the New Jersey Turnpike. The NJ Turnpike's signs of that period, and up to around 2012, were capable of displaying a few messages in neon, all oriented around warning drivers to slow down: "REDUCE SPEED", followed by a warning of either construction, accident, congestion, ice, snow, or fog at a certain distance ahead. The New Jersey Turnpike Authority replaced those signs (along with 1990s-vintage dot-matrix VMS's along the Garden State Parkway) with more flexible electronic signs between 2010 and 2016.

The current VMS systems are largely deployed on freeways, trunk highways, or in work zones.

On the interchange of I-5 and SR 120 in San Joaquin County, California, an automated visibility and speed warning system was installed in 1996 to warn traffic of reduced visibility due to fog (where tule fog is a common problem in the winter), and of slow or stopped traffic.

Message Signs were deployed in Ontario during the 1990s in the Greater Toronto Area and are now being upgraded on 400 series Highways in the GTA as well as two pilot secondary highways in northeastern Ontario.

Technologies and Types

Early variable message signs included static signs with words that would illuminate (often using neon tubing) indicating the type of incident that occurred, or signs that used rotating prisms (trilons) to change the message being displayed. These were later replaced by dot matrix displays typically using eggcrate, fiber optic, or flip-disc technology, which were capable of displaying a much wider range of messages than earlier static variable message signs. Since the late 1990s, the most common technology used in new installations for variable message signs are LED displays. In recent years, some newer LED variable message signs have the ability to display colored text and graphics.

Dot-matrix variable message signs are divided into three subgroups: character matrix, row matrix, and full matrix. In a character matrix VMS, each character is given its own matrix with equal horizontal spacing between them, typically with two or three rows of characters. In a full matrix VMS, the entire sign is a single large dot matrix display, allowing the display of different fonts and graphics. A row matrix VMS is a hybrid of the two types, divided into two or three rows like a character matrix display, except each row is a single long dot matrix display instead of being split per character horizontally.

Overhead variable message signs are today available in three form factors: front access, rear access, and walk-in. In a front access variable message sign, maintenance is performed by lifting the sign open from the front. Most smaller VMS are of the front access form factor, and are typically installed today on major arterials. The rear access form factor is similar to the front access form factor, except that maintenance is performed from the rear of the sign, and are commonly used for medium-sized dynamic message signs installed along the roadside of freeways (instead of overhead). The walk-in form factor is a more recent introduction, where maintenance on the sign is performed from the inside of the sign. A key advantage of the walk-in form factor is that lane closures are generally not required to perform maintenance on the sign. Most of the largest VMS units installed today are walk-in units, and are typically installed overhead on freeways.

The NJ Turnpike Authority counts five unique types of variable message signs under its jurisdiction, at least one of which has been replaced by newer signs. They are:

1. "REDUCE SPEED" neon signs (1950s-2010, obsolete, have been replaced).
2. "Changeable message signs" (trilon/ rotating-drum signs that can be used for closing roads or moving traffic to other roadways).
3. Electronic VMS: signs with remotely controlled messages displayed on them; the messages are sent from the State Traffic Management Center, updating the signs automatically.
4. Variable speed limit signs - used for varying the posted speed limits within work zones and in emergencies.
5. Portable VMS: movable "electronic VMS". A portable VMS has much the same characteristics as a fixed electronic VMS, but can be moved from location to location as the need dictates.

Usage

Early models required an operator to be physically present when programming a message, whereas newer models may be reprogrammed remotely via a wired or wireless network or cellphone connection.

A complete message on a panel generally includes a problem statement indicating incident, roadwork, stalled vehicle etc.; a location statement indicating where the incident is located; an effect statement indicating lane closure, delay, etc. and an action statement giving suggestion what to do traffic conditions ahead. These signs are also used for AMBER Alert and Silver Alert messages.

In some places, VMSes are set up with permanent, semi-static displays indicating predicted travel times to important traffic destinations such as major cities or interchanges along the route of a highway.

Typical messages provide the following information:

- Promotional messages about services provided by a road authority during non-critical hours, such as carpooling efforts, travelers' information stations and 5-1-1 lines
- Crashes, including vehicle spin-out or rollover
- Road Works
- Incidents affecting normal traffic flow in a lane or on shoulders
- Non-recurring congestion, often a residual effect of cleared crash
- Closures of an entire road, e.g. over a mountain pass in winter.
- Exit ramp closures
- Debris on roadway
- Vehicle fires
- Short-term maintenance or construction lasting less than three days

- Pavement failure alerts
- AMBER Alerts and weather warnings via the warning infrastructure of NOAA Weather Radio's SAME system
- Travel times
- Variable speed limits
- Car park occupancy levels

The information comes from a variety of traffic monitoring and surveillance systems. It is expected that by providing real-time information on special events on the oncoming road, VMS can improve motorists' route selection, reduce travel time, mitigate the severity and duration of incidents and improve the performance of the transportation network.

Movable Versions

Trailer-mounted VMS in central London, England

Truck-mounted VMSes are sometimes dispatched by highway agencies such as Caltrans to warn traffic of incidents such as accidents in areas where permanent VMSes aren't available or near enough as a preventive measure for reducing secondary accidents. They are often deployed in pairs so that the second VMS truck can take over when the traffic queue overtakes the first truck, requiring the first truck to reposition further upstream from the queue, to be effective. An optional third truck, the team leader, may be utilized for driving by and monitoring the incident itself, traffic patterns and delay times, to make strategic decisions for minimizing delays—analogous to spotter planes used in fighting forest fires.

Trailer-mounted variable-message signs are used to alter traffic patterns near work zones, and for traffic management for special events, natural disasters, and other temporary traffic patterns. The messages displayed on the sign can be programmed locally on the unit's control panel, or units equipped with a cellular modem can be programmed remotely via computer or phone. Most manufacturers produce trailers which comply with the National Transportation Communications for Intelligent Transportation System Protocol (NTCIP) which allows the portable trailer to be integrated with an intelligent transportation system. Trailer-mounted VMS can be equipped with radar, cameras, and other sensing devices as part of a smart work zone deployment..

In Popular Culture

A variable-message sign figures significantly into the plot of the 1991 film *L.A. Story*.

Cycling Infrastructure

Cycling infrastructure being placed in Chicago, Illinois.

Signposted greenway, bordering on a gracht in Nordhorn, Germany

Cyclists use a segregated cut through of a busy interchange in London at rush hour.

Cycling infrastructure refers to all infrastructure which may be used by cyclists. This includes the same network of roads and streets used by motorists, except those roads from which cyclists have been banned (e.g., many freeways/motorways), plus additional bikeways that are not available to motor vehicles, such as bike paths, bike lanes, cycle tracks and, where permitted, sidewalks, plus amenities like bike racks for parking and specialized traffic signs and signals.

The manner in which the public road network is designed, built and managed can have a significant effect on the utility and safety of cycling. The cycling network may be able to provide the users with direct, convenient routes minimizing unnecessary delay and effort in reaching their destinations. Settlements with a dense road network of interconnected streets tend to be viable utility cycling environments.

History

Bikeways

A bikeway is a lane, route, way or path which in some manner is specifically designed and /or designated for bicycle travel. Bike lanes demarcated by a painted marking are quite common in many cities. Cycle tracks demarcated by barriers, bollards or boulevards are quite common in some European countries such as the Netherlands, Denmark and Germany. They are also increasingly common in other major cities such as New York City, Melbourne, Ottawa, Vancouver and San Francisco. Montreal and Davis, California, which have had segregated cycling facilities with barriers for several decades, are among the earliest examples in North American cities.

Terminology

Segregated cycle facility in Karlsruhe, Germany. Fahrradstrasse means "bicycle street."

Some bikeways are separated from motor traffic by physical constraints (e.g. barriers, parking or bollards)—bicycle trail, cycle track—but others are partially separated only by painted markings—bike lane, buffered bike lane, and contraflow bike lane. Some share the roadway with motor vehicles—bicycle boulevard, sharrow, advisory bike lane—or shared with pedestrians—greenway, shared use path.

Segregation

The term bikeway is largely used in North America to describe all routes that have been designed or updated to encourage more cycling or make cycling safer. In some jurisdictions such as the

United Kingdom, *segregated cycling facility* is sometimes preferred to describe cycling infrastructure which has varying degrees of separation from motorized traffic, or which has excluded pedestrian traffic in the case of exclusive bike paths.

There is no single usage of *segregation*; in some cases it can mean the exclusion of motor vehicles and in other cases the exclusion of pedestrians as well. Thus, it includes bike lanes with solid painted lines but not lanes with dotted lines and advisory bike lanes where motor vehicles are allowed to encroach on the lane. It includes cycle tracks as physically distinct from the roadway and sidewalk (e.g. barriers, parking or bollards). And it includes bike paths in their own right of way exclusive to cycling. Paths which are shared with pedestrians and other non-motorized traffic are *not* considered segregated and are typically called shared use path, multi-use path in North America and shared-use footway in the UK.

Guides

Various guides exist to define the different types of bikeway infrastructure, including UK Department for Transport manual *The Geometric Design of Pedestrian, Cycle and Equestrian Routes*, Sustrans Design Manual, UK Department of Transport Local Transport Note 2/08: Cycle infrastructure design the Danish Road Authority guide *Registration and classification of paths*, the Dutch *CROW*, the American Association of State Highway and Transportation Officials (AASHTO) *Guide to Bikeway Facilities*, the Federal Highway Administration (FHWA) *Manual on Uniform Traffic Control Devices* (MUTCD), and the US National Association of City Transportation Officials (NACTO) *Urban Bikeway Design Guide*.

Safety

There have been a lot of studies on the safety of all types of bikeways. Proponents say that segregation of cyclists from fast or frequent motorized traffic is necessary to provide a safe and welcoming cycling environment. Opponents point out the increased risk from various types of infrastructure including shared use paths.

Legislation

Different countries have different ways to legally define and enforce bikeways.

Bikeway Controversies

Controversies have surrounded bikeways, particularly in North America and the United Kingdom. Some detractors argue that one must be careful in interpreting the operation of dedicated or segregated bikeways/cycle facilities across different designs and contexts; what works for the Netherlands won't necessarily work elsewhere. Proponents argue that dedicated bike lanes have been implemented in many cities and are both popular and safe.

Bikeway Selection

Jurisdictions have guidelines around the selection of the right bikeway treatments in order make routes more comfortable and safer for cycling.

Bikeway Types

Bikeways can fall into these main categories: separated in-roadway bikeways such as bike lanes and buffered bike lanes; physically separated in-roadway bikeways such as cycle tracks; right-of-way paths such as bike paths and shared use paths; and shared in-roadway bikeways such as bike boulevards, shared lane markings, and advisory bike lanes. The exact categorization changes depending on the jurisdiction and organization, while many just list the types by their commonly used names

Separated in-roadway Bikeway

Bike Lanes

Bike lanes, or cycle lanes, are on-road lanes marked with paint dedicated to cycling and typically excluding all motorized traffic.

Buffered Bike Lanes

A buffered bike lane is typically a lane with a wide painted buffer to demarcate a larger gap between the cycle lane and other traffic.

Contraflow Bike Lanes

Contraflow bike lanes are typically a painted lane added to some one-way streets as a way to allow cycling traffic to safely travel in the opposite direction of all other traffic, including motorized vehicles and bicycles.

Physically Separated in-roadway Bikeway

Cycle Tracks / Separated Bike Lanes / Protected Bike Lanes

Extremely broad separated bike lanes on Dronning Louises Bro in Copenhagen, Denmark. Normally filled with bicycles, as it has been the busiest stretch of bicycle lane in the world. The title have been taken over by Knippelsbro, another bridge in Copenhagen.

A physically marked and separated lane dedicated for cycling that is on or directly adjacent to the roadway but typically excludes all motorized traffic with some sort of vertical barrier.

Bike Paths

Bike paths are paths with their own right of way dedicated to cycling, though in many cases shared with pedestrians and other non-motorized traffic.

Greenways

Former railroad line transformed into bicycle path between Metelen and Steinfurt in Germany

A greenway is a long, narrow piece of land, often used for recreation and pedestrian and bicycle user traffic, and sometimes for streetcar, light rail or retail uses.

Shared-use Path

A shared use path supports multiple modes, such as walking, bicycling, inline skating and people in wheelchairs.

Shared in-roadway Bikeways

Bicycle Boulevards

A bicycle boulevard is a low speed street which has been optimized for bicycle traffic. Bicycle boulevards discourage cut-through motor vehicle traffic but allow local motor vehicle traffic. They are designed to give priority to cyclists as through-going traffic.

Shared Lane Markings

Advisory Bike Lane

An advisory bike lane is a bike lane which motorists may legally encroach. Advisory bike lanes are often installed on low volume streets and in combination with a center line removal. These markings give bicyclists a space to ride, but are also available to motorists if space is needed to pass oncoming traffic.

Roadways with Legal Access for Cycling

Bicyclists are legally allowed to travel on many roadways in accordance with the rules of the road for drivers of vehicles.

Road Shoulder

A road shoulder is a reserved lane on the verge of a roadway that is often used by bicyclists and also serves as an emergency stopping lane for motor vehicles.

Bicycle Superhighways

Denmark and the Netherlands have pioneered the concept of "bicycle superhighways" to increase the speed, safety, and comfort of bicycle commuting. Since then London (UK) and Flanders have also started building bicycle superhighways. The physical design of the routes changes depending on the circumstances and can include shared in-road bikeways, bike lanes and cycle tracks, making it difficult to categorize into degree of separation from other traffic.

The first Dutch route opened in 2004 between Breda and Etten-Leur; many others have been added since then. The first Danish route, C99, opened in 2012 between the Vesterbro rail station in Copenhagen and Albertslund, a western suburb. The route cost 13.4 million DKK and is 17.5 km long, built with few stops and new paths away from traffic. "Service stations" with air pumps are located at regular intervals, and where the route must cross streets, handholds and running boards are provided so cyclists can wait without having to put their feet on the ground.

The cost of building a bicycle super highway depends on many things, but is usually between €300000/km (for a wide dedicated cycle track) and €800000/km (when complex civil engineering structures are needed).

Cycling-friendly Streetscape Modifications

There are various measures cities and regions often take on the roadway to make it more cycling friendly and safer. Aspects of infrastructure may be viewed as either cyclist-hostile or as cyclist-friendly. Measures to encourage cycling include traffic calming; traffic reduction; junction treatment; traffic control systems to recognize cyclists and give them priority; exempt cyclists from banned turns and access restrictions; implement contra-flow cycle lanes on one-way streets; implement on-street parking restrictions; provide advanced stop lines/bypasses for cyclists at traffic signals; marking wide curb/kerb lanes; and marking shared bus/cycle lanes.

Traffic Reduction

Removing traffic can be achieved by straightforward diversion or alternatively reduction. Diversion involves routing through-traffic away from roads used by high numbers of cyclists and pedestrians. Examples of diversion include the construction of arterial bypasses and ring roads around urban centres.

Indirect methods involve reducing the infrastructural capacity dedicated to moving motorized vehicles. This can involve reducing the number of road lanes, closing bridges to certain vehicle types and creating vehicle restricted zones or environmental traffic cells. In the 1970s the Dutch city of Delft began restricting private car traffic from crossing the city centre. Similarly, Groningen is divided into four zones that cannot be crossed by private motor-traffic, (private cars must use the ring road instead). Cyclists and other traffic can pass between the zones and cycling accounts for 50%+ of trips in Groningen (which reputedly has the third highest proportion of cycle traffic of any city). The Swedish city of Gothenburg uses a similar system of traffic cells.

Another approach is to reduce the capacity to park cars. Starting in the 1970s, the city of Copenhagen, which is now noted for high cycling levels, adopted a policy of reducing available car parking capacity by several per cent per year. The city of Amsterdam, where around 40% of all trips are by bicycle, adopted similar parking reduction policies in the 80s and 90s.

Direct traffic reduction methods can involve straightforward bans or more subtle methods like road pricing schemes or road diets. The London congestion charge reportedly resulted in a significant increase in cycle use within the affected area.

Traffic Calming

Speed reduction has traditionally been attempted by statutory speed limits.

Recent implementations of shared space schemes have delivered significant traffic speed reductions. The reductions are sustainable, without the need for speed limits or speed limit enforcement. In Norrköping, mean traffic speeds in 2006 dropped from 21 to 16 km/h (13 to 10 mph) since the implementation of such a scheme.

Even without shared street implementation, creating 30 km/h zones (or 20 mph zone) has been shown to reduce crash rates and increase numbers of cyclists and pedestrians. Other studies have revealed that lower speeds reduce community severance caused by high speed roads. Research has shown that there is more neighborhood interaction and community cohesion when speeds are reduced to 20 mph.

One-way Streets

German research indicates that making one-way streets two-way for cyclists results in a reduction in the total number of collisions. In Belgium, all one-way streets in 50 km/h zones are by default two-way for cyclists. A Danish road directorate states that in town centres it is important to be able to cycle in both directions in all streets, and that in certain circumstances, two-way cycle traffic can be accommodated in an otherwise one-way street.

Two-way Cycling on One-way Streets

One-way street systems can be viewed as either a product of traffic management that focuses on trying to keep motorized vehicles moving regardless of the social and other impacts, such as by some cycling campaigners, or seen as a useful tool for traffic calming, and for eliminating rat runs, in the view of UK traffic planners.

One-way streets can disadvantage cyclists by increasing trip-length, delays and hazards associated with weaving maneuvers at junctions. In northern Europe countries such as the Netherlands, however, cyclists are frequently granted exemptions from one-way street restrictions, which improves cycling traffic flow while restricting motorized vehicles.

German research indicates that making one-way streets two-way for cyclists results in a reduction in the total number of collisions.

There are often restrictions to what one-way streets are good candidates for allowing two-way cycling traffic. In Belgium road authorities can in principle allow any one-way streets in 50 kilometres per hour (31 mph) zones to be two-way for cyclists if the available lane is at least 3 metres (9.8 ft) wide (area free from parking) and no specific local circumstances prevent it. Denmark, a country with high cycling levels, does not use one-way systems to improve traffic flow. Some commentators argue that the initial goal should be to dismantle large one-way street systems as a traffic calming/traffic reduction measure, followed by the provision of two-way cyclist access on any one-way streets that remain.

Intersection/Junction Design

In general, junction designs that favour higher-speed turning, weaving and merging movements by motorists tend to be hostile for cyclists. Free-flowing arrangements can be hazardous for cyclists and should be avoided. Features such as large entry curvature, slip-roads and high flow roundabouts are associated with increased risk of car–cyclist collisions. Cycling advocates argue for modifications and alternative junction types that resolve these issues such as reducing kerb radii on street corners, eliminating slip roads and replacing large roundabouts with signalized intersections.

Protected Intersection

Another approach which the Netherlands innovated is called in North America protected intersection which reconfigures intersections to reduce risk to cyclists as they cross or turn. Some American cities are starting to pilot protected intersections.

Bike Box

A bike box is a designated area at the head of a traffic lane at a signalized intersection that provides bicyclists with a safe and visible way to get ahead of queuing traffic during the red signal phase.

Roundabouts

On large roundabouts of the design typically used in the UK and Ireland, cyclists have an injury accident rate that is 14-16 times that of motorists. Research indicates that excessive sight lines at uncontrolled intersections compound these effects. In the UK, a survey of over 8,000 highly experienced and mainly adult male Cyclists Touring Club members found that 28% avoided roundabouts on their regular journey if at all possible. The Dutch CROW guidelines recommend roundabouts only for intersections with motorized traffic up to 1500 per hour. In order to accommodate greater volumes of traffic they recommend traffic light intersections or grade separation for cyclists. Examples of grade separation for cyclists include tunnels, or more spectacularly, raised "floating" roundabouts for cyclists.

Traffic Signals/Traffic Control Systems

A bicycle signal light in Toronto

How traffic signals are designed and implemented directly impacts cyclists. For instance, poorly adjusted vehicle detector systems, used to trigger signal changes, may not correctly detect cyclists. This can leave cyclists in the position of having to "run" red lights if no motorized vehicle arrives to trigger a signal change. Some cities use urban adaptive traffic control systems (UTCs), which use linked traffic signals to manage traffic in response to changes in demand. There is an argument that using a UTC system merely to provide for increased capacity for motor traffic will simply drive growth in such traffic. However, there are more direct negative impacts. For instance, where signals are arranged to provide motor traffic with so called green waves, this can create "red waves" for other road users such as cyclists and public transport services. Traffic managers in Copenhagen have now turned this approach on its head and are linking cyclist-specific traffic signals on a major arterial bike lane to provide green waves for rush hour cycle-traffic. However, this would still not resolve the problem of red-waves for slow (old and young) and fast (above average fitness) cyclists. Cycling-specific measures that can be applied at traffic signals include the use of advanced stop lines and/or bypasses. In some cases cyclists might be given a free-turn or a signal bypass if turning into a road on the nearside.

Signposting

One of the mountain pass cycling milestones placed along the climb to the Col d'Izoard in the French Alps.

In many places worldwide special signposts for bicycles are used to indicate directions and distances to destinations for cyclists. Apart from signposting in and between urban areas, mountain pass cycling milestones have become an important service for bicycle tourists. They provide cyclists with information about their current position with regard to the summit of the mountain pass.

Widening Outside Lanes

A bike lane in North America.

One method for reducing potential friction between cyclists and motorized vehicles is to provide "wide kerb", or "nearside", lanes (UK terminology) or "wide outside through lane" (U.S. terminology). These extra-wide lanes increase the probability that motorists pass cyclists at a safe distance without having to change lanes. This is held to be particularly important on routes with a high proportion of wide vehicles such as buses or heavy goods vehicles (HGVs). They also provide more room for cyclists to filter past queues of cars in congested conditions and to safely overtake each other. Due to the tendency of all vehicle users to stay in the center of their lane, it would be necessary to sub-divide the cycle lane with a broken white line to facilitate safe overtaking. Overtaking is indispensable for cyclists, as speeds are not dependent on the legal speed limit, but on the rider's capability.

The use of such lanes is specifically endorsed by *Cycling: the way ahead for towns and cities*, the European Commission policy document on cycle promotion. The UK Cyclists Touring Club argues for a marked lane width of 4.25 m (14 ft). It is argued that, on undivided roads, this width provides cyclists with adequate clearance from passing wide vehicles while being sufficiently narrow to deter motorists from attempting to "double up" and form two lanes. This "doubling up" effect may be related to junctions. At non-junction locations, greater width might be preferable if this effect can be avoided.

Shared Space

Shared space schemes extend this principle further by removing the reliance on lane markings altogether, and also removing road signs and signals, allowing all road users to use any part of the road, and giving all road users equal priority and equal responsibility for each other's safety. Experiences where these schemes are in use show that road users, particularly motorists, undirected by signs, kerbs, or road markings, reduce their speed and establish eye contact with other users.

Results from the thousands of such implementations worldwide all show casualty reductions and most also show reduced journey times. After the partial conversion of London's Kensington High Street to shared space, accidents decreased by 44% (the London average was 17%).

New Road, Brighton - Shared Space scheme reduced motor traffic by 93%.

CFI argues for a marked lane width of 4.25 metres (13.9 ft). On undivided roads, width provides cyclists with adequate clearance from passing HGVs while being narrow enough to deter drivers from "doubling up" to form two lanes. This "doubling up" effect may be related to junctions. At non-junction locations, greater width might be preferable if this effect can be avoided. The European Commission specifically endorses wide lanes in its policy document on cycling promotion, *Cycling: the way ahead for towns and cities.*

A bus and cycle lane in Mannheim, Germany

Shared Bus and Cycle Lanes

Shared bus and cycle lane|Shared bus and cycle lanes are also a widely endorsed method for providing a more comfortable and safer space for cyclists. Depending on the width of the lane, the speeds and number of buses, and other local factors, the safety and popularity of this arrangement vary.

Road Surface

Bicycle tires being narrow, road surface is more important than for other transport, for both comfort and safety. The type and placement of storm drains, manholes, surface markings, and the general road surface quality should all be taken into account by a bicycle transportation engineer. Drain grates, for example, must not catch wheels.

Trip-end Facilities

Bicycle Parking/Storage Arrangements

Bicycle parking at the Alewife subway station in Cambridge, Massachusetts, located at the intersection of three cycle paths.

Bicycle parking lot in Amsterdam.

As secure and convenient bicycle parking is a key factor in influencing a person's decision to cycle, decent parking infrastructure must be provided to encourage the uptake of cycling. Decent bicycle parking involves weather-proof infrastructure such as lockers, stands, manned or unmanned bicycle parks, as well as bike parking facilities within workplaces to facilitate bicycle commuting. It also will help if certain legal arrangements are put into place to enable legitimate *ad hoc* parking, for example to allow people to lock their bicycles to railings, signs and other street furniture when individual proper bike stands are unavailable.

Other Trip end Facilities

Some people need to wear special clothes such as business suits or uniforms in their daily work.

In some cases the nature of the cycling infrastructure and the prevailing weather conditions may make it very hard to both cycle and maintain the work clothes in a presentable condition. It is argued that such workers can be encouraged to cycle by providing lockers, changing rooms and shower facilities where they can change before starting work.

Theft Reduction Measures

The theft of bicycles is one of the major problems that slow the development of urban cycling. Bicycle theft discourages regular cyclists from buying new bicycles, as well as putting off people who might want to invest in a bicycle.

Several measures can help reduce bicycle theft:

- Bicycle registration to enable recovery if stolen
- Making cyclists aware of antitheft devices and their effective use
- Mounting sting operations to catch thieves
- Secure bicycle parking: offering safe bicycle parking facilities such as guarded bicycle parking (manned or with camera surveillance) or bicycle lockers
- Promoting devices to enable remote tracking of a bicycle's location
- Targeting cycle thieves
- Using folding bicycles which can be safely stored (for example) in cloakrooms or under desks.

Certain European countries apply such measures with success, such as the Netherlands or certain German cities using registration and recovery. Since mid-2004, France has instituted a system of registration, in some places allowing stolen bicycles to be put on file in partnership with the urban cyclists' associations. This approach has reputedly increased the stolen bicycle recovery rate to more than 40%. By comparison, before the commencement of registration, the recovery rate in France was about 2%.

In some areas of the United Kingdom, bicycles fitted with location tracking devices are left poorly secured in theft hot-spots. When the bike is stolen, the police can locate it and arrest the thieves. This sometimes leads to the dismantling of organized bicycle theft rings, as bike theft generally enjoys a very low priority with the police.

Bicycle Lift

Some cyclists have difficulty climbing steep hills, and devices such as the Trampe bicycle lift, in Trondheim, have been developed to help alleviate this problem.

Integration with Public Transit

Cycling can often be integrated successfully with other transport modes. For example, in the Netherlands and Denmark a large number of train journeys may start by bicycle. In 1991, 44% of Dutch train travelers went to their local station by bicycle and 14% used a bicycle at their destinations. The key ingredients for this are claimed to be:

- an efficient, attractive and affordable train service
- secure bike parking at train stations
- a quick and easy bicycle rental system for commuters, the OV-bicycle scheme, at train stations
- a town planning policy that results in a sufficient proportion of the potential commuter population (e.g. 44%) living/working within a reasonable cycling distance of the train stations.

It has been argued in relation to this aspect of Dutch or Danish policy that ongoing investment in rail services is vital to maintaining their levels of cycle use.

Bike commuters disembark at Palo Alto Station in Palo Alto, California

An often forgotten major success story is the integration of cycling and public transport is Japan. Starting in 1978, Japan expanded bicycle parking supply at railway stations from 598,000 spaces in 1977 to 2,382,000 spaces in 1987. As of 1987, Japanese provisions included 516 multi-story garages for bicycle parking.

In January 2007, the European parliament adopted a motion decreeing that all international trains must carry bicycles. In some cities, bicycles may also be carried on local trains, trams and buses so that they may be used at either end of the trip. The Rheinbahn transit company in Düsseldorf permits bicycle carriage on all its bus, tram and train services at any time of the day,. In Munich bicycles are allowed on the S-Bahn commuter trains outside of rush hours, and folding bikes are allowed on city busses. In Copenhagen, you can take your bicycle with you in the S-tog commuter trains, all times a day with no additional costs. In France, the prestigious TGV high-speed trains are even having some of their first class capacity converted to store bicycles. There have also been schemes, such as in Victoria, British Columbia, Acadia, and Canberra, Australia to provide bicycle carriage on buses using externally mounted bike carriers.

In Edmonton, Alberta, Canada, busses on most city routes have externally mounted carriers for bicycles, and bikes are allowed on the light rail trains at no extra cost outside of rush hour. All public transit buses in Chicago and suburbs allow up to two bikes at all times. The same is true of Grand River Transit buses in the Region of Waterloo, Ontario, Canada. Trains allow bikes with some restrictions. Where such services are not available, some cyclists get around this restriction by removing their pedals and loosening their handlebars as to fit into a box or by using folding bikes that can be brought onto the train or bus like a piece of luggage. The article on buses in Christchurch, New Zealand lists 27 routes with bike racks.

However, there are strong cultural variations in how cycling is treated in such situations. For instance in the Irish university city of Galway, the same town that suggested cyclists dismount and walk across each intersection, the secure parking of bikes is forbidden within the grounds of the central train station. However, cut-price car parking is available for motorists holding a valid train ticket.

Bikesharing Systems

A bicycle sharing system, public bicycle system, or bike share scheme, is a service in which bicycles are made available for shared use to individuals on a very short term basis. Bike share schemes allow people to borrow a bike from point "A" and return it at point "B". Many of the bicycle sharing systems are on a subscription basis.

Examples of Cycling Infrastructure

Cycling in Santos, Brazil

Bikeway in Portugal

Bikeway in Pocuro, Chile

Ciclopaseo, Ciclovia Ejido Park Quito, Ecuador

Pedestrian Crossing

This crosswalk is used by students at Purchase College, New York.

Zebra crossing in Edinburgh, Scotland, UK

A pedestrian crossing or crosswalk is a place designated for pedestrians to cross a road. Crosswalks are designed to keep pedestrians together where they can be seen by motorists, and where they can cross most safely across the flow of vehicular traffic.

Marked pedestrian crossings are often found at intersections, but may also be at other points on busy roads that would otherwise be too unsafe to cross without assistance due to vehicle numbers, speed or road widths. They are also commonly installed where large numbers of pedestrians are attempting to cross (such as in shopping areas) or where vulnerable road users (such as school children) regularly cross. Rules govern usage of the pedestrian crossings to ensure safety; for example, in some areas, the pedestrian must be more than halfway across the crosswalk before the driver proceeds.

Signalised pedestrian crossings clearly separate when each type of traffic (pedestrians or road vehicles) can use the crossing. Unsignalized crossings generally assist pedestrians, and usually prioritise pedestrians, depending on the locality. What appear to be just pedestrian crossings can also be created largely as a traffic calming technique, especially when combined with other features like pedestrian priority, refuge islands, or raised surfaces.

History

Pedestrian crossings were common in the Roman city of Pompeii.

Pedestrian crossings already existed more than 2000 years ago, as can be seen in the ruins of Pompeii. Blocks raised on the road allowed pedestrians to cross the street without having to step onto the road itself which doubled up as Pompeii's drainage and sewage disposal system. The spaces between the blocks allowed horse-drawn carts to pass along the road.

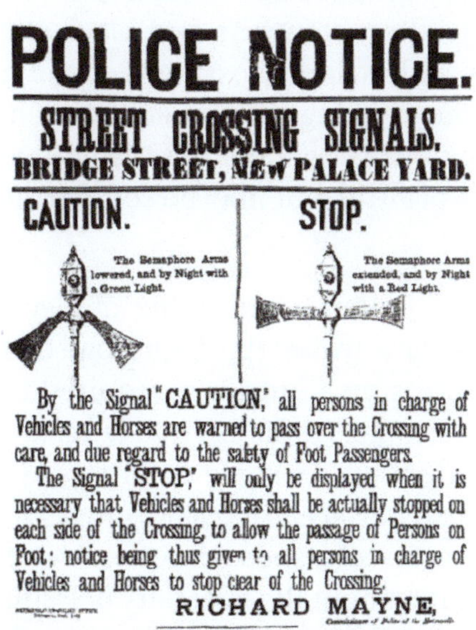

POLICE NOTICE.

STREET CROSSING SIGNALS.

BRIDGE STREET, NEW PALACE YARD.

CAUTION.

The Semaphore Arms lowered, and by Night with a Green Light.

STOP.

The Semaphore Arms extended, and by Night with a Red Light.

By the Signal "CAUTION," all persons in charge of Vehicles and Horses are warned to pass over the Crossing with care, and due regard to the safety of Foot Passengers.

The Signal "STOP," will only be displayed when it is necessary that Vehicles and Horses shall be actually stopped on each side of the Crossing, to allow the passage of Persons on Foot; notice being thus given to all persons in charge of Vehicles and Horses to stop clear of the Crossing.

RICHARD MAYNE,

Police notice explaining the operation of the first pedestrian crossing signal.

The first pedestrian crossing signal was erected in Bridge Street, Westminster, London in December 1868. It was the idea of John Peak Knight, a railway engineer, who thought that it would provide a means to safely allow pedestrians to cross this busy thoroughfare. The signal consisted of a semaphore arm (Manufactured by Saxby and Farmer, who were railway signaling makers), which was raised and lowered manually by a police constable who would rotate a handle on the side of the pole. The semaphore arms were augmented by gas illuminated lights at the top (green and red) to increase visibility of the signal at night. However, in January 1869, the gas used to illuminate the lights at the top leaked and caused an explosion, injuring the police operator. No further work was done on signalled pedestrian crossings until fifty years later.

In the early days of the 20th century car traffic increased dramatically. A reader of The Times wrote to the editor in 1911:

"Could you do something to help the pedestrian to recover the old margin of safety on our common streets and roads? It is heartrending to read of the fearful deaths taking place. If a pedestrian now has even one hesitation or failure the chance of escape from a dreadful death is now much less than when all vehilcles were much slower. There is, too, in the motor traffic an evident desire not to slow down before the last moment. It is surely a scandal that on the common ways there should be undue apprehension in the minds of the weakest users of them. While the streets and roads are for all, of necessity the pedestrians, and the feeblest of these, should receive the supreme consideration."

Types

Crossings are of various types.

- Unmarked crossings may occur at any intersection. In the US these are called "unmarked crosswalks."
- The simplest marked crossings may just consist of some markings on the road surface. In the US these are known as "marked crosswalks." In the UK these are often called Zebra crossings, referring to the alternate white and black stripes painted on the road surface. Depending on local laws, pedestrians crossing the road may or may not have priority over road traffic when using the crossing. If the pedestrian has priority, then they have an incentive to use the crossing instead of crossing the road at other places. In some countries, pedestrians may not have priority, but may be committing an offence if they cross the road elsewhere.
- Some crossings have special signals consisting of electric lamps or light-emitting diode (LED) panels. The signals allow pedestrians and road traffic to use the crossing alternately. On some traffic signals, pressing a button is required to trigger the signal. These signals may be integrated into a regular traffic light arrangement or may be on their own if the crossing is not at an intersection. In some countries approach radar or electro-magnetic detectors buried in the road surface are used to measure the speed of oncoming vehicles and used to set the lights to red if a speeding offence is detected. This has the effect of enforcing the local speed limit without the necessity of issuing speeding citations, etc. Audible or tactile signals may also be included to assist people who have poor sight. In many cities,

some or most signals are equipped with countdown timers to give notice to both drivers and pedestrians the time remaining on the crossing signal.

- Sites with extremely high traffic, or roads where pedestrians are not allowed (freeways or motorways) may instead be crossed via pedestrian bridges or tunnels. A variation on the bridge concept, often called a skyway or skywalk, is sometimes implemented in regions that experience inclement weather.

Special markings are often made on the road surface, both to direct pedestrians and to prevent motorists from stopping vehicles in the way of foot traffic. There are many varieties of signal and marking layouts around the world and even within single countries. In the United States, there are many inconsistencies, although the variations are usually minor. There are several distinct types in the United Kingdom, each with their own name.

Pedestrian refuges or small islands in the middle of a street may be added when a street is very wide, as these crossings can be too long for some individuals to cross in one cycle. In places where there is very high pedestrian traffic, exclusive traffic signal phases for pedestrians (also known as Barnes Dances) may be used, which stop vehicular traffic in all directions at the same time. Another relatively widespread variation is the curb/kerb extension (also known as a bulb-out) which narrows the width of the street and is used in combination with crosswalk markings.

Distinctions by Region

North America

In the United States, crosswalks are sometimes marked with white stripes, though many municipalities have slightly different methods, styles, or patterns for doing so, and the styles may vary over time as intersections are built and reconstructed. There are two main methods for road markings in the United States. Most frequently, they are marked with two thick white lines running from one side of the road to the other. A third "stop line", which is very thick and extends only across lanes going into the intersection, is usually also present. Left-turn stop lines are often set further back, to avoid conflict with left-turning traffic coming from the roadway on the right. The stop line acts as the legally mandated stopping point for vehicles, and discourages drivers from stopping in the middle of the crosswalk. The other method involves the use of the more easily visible continental stripes (like UK zebra crossings), which are becoming more popular in place of the two-line variant. The designs used vary widely between jurisdictions, and often vary even between a city and its county (or local equivalents). Where a road forms part of a city limit or other such political boundary—thus making the intersection shared between the two—there may be more than one design used on different sides, depending upon which government painted it.

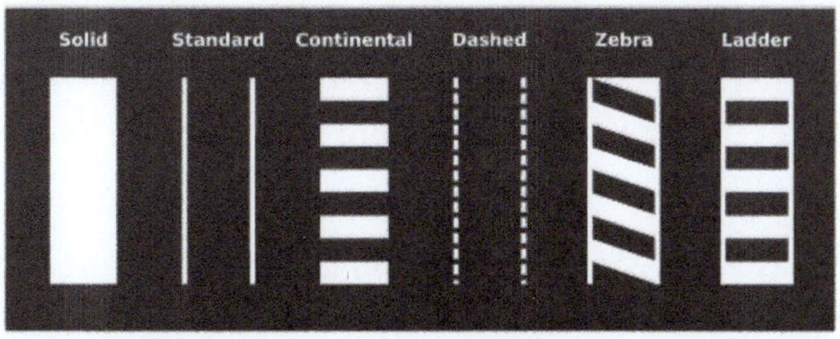

Crosswalk pavement marking variants per the U.S. FHWA.

Marked crosswalks are usually placed at traffic intersections or crossroads, but are occasionally used at mid-block locations where pedestrian generators are present such as at transit stops, schools, retail, or housing destinations. In the United States, these so-called "mid-block" crossings may include additional regulatory signage such as "PED XING" (for "pedestrian crossing"), flashing yellow beacons, stop or yield signs, or by actuated or automatic signals. Some more innovative crossing treatments include in-pavement flashers, yellow flashing warning lights installed in the roadway, or HAWK beacon, an overhead signal with two pair of red beacons above an amber beacon, when a pedestrian is detected or actuates the device it begins a sequence of amber flashing followed by a solid red [when vehicles may not cross], followed by a flashing red phase that allows motorists to proceed, only if the pedestrian[s] are clear of the travel way. In the United States, crossing laws vary from state to state and sometimes at the local level. All states require vehicles to yield to a pedestrian who has entered a marked crosswalk. To gain the right-of-way in some parts of Canada, however, the pedestrian holds out his hand in a position much like that used to shake hands, and steps off the curb.

A sign warning motorists to yield to those crossing the crosswalk.

At crossings controlled by signals, the most common variety is arranged like this: At each end of a crosswalk, the poles which hold the traffic lights also have white "walk" and Portland Orange "don't walk" signs. These particular colors are used in North America (excluding Quebec) to provide conspicuity against the backdrop of red, yellow, and green traffic lights. Modern signals generally use pictograms of an orange upraised hand and a white walking pedestrian rather than words. As a warning, the "don't walk" or hand signals begin to blink when the transition to "don't walk" is imminent. This normally occurs several seconds before the light turns yellow, usually going solid orange when the traffic light turns yellow. Some signals continue flashing the hand/"don't walk" phase during the yellow light, and go steady at red. Sometimes the "walk" signal does not come on in a steady pattern; it will sometimes blink/flash on and off instead to warn pedestrians to cross the street with caution due to the possibility of a turning vehicle. On pedestrian signals displaying text, "don't walk" is spelled without an apostrophe so that it fits easily on the sign. A black baffle is customarily placed in front of the lights to shield them from the sun and increase their visibility, as well as protect them from damage.

Legally speaking, in most states crosswalks exist at all intersections meeting at approximately right angles, whether they are marked or not. All states except Maine and Michigan require vehicles to yield to a pedestrian who has entered an unmarked crosswalk. In some cities in the US, other methods of pedestrian detection are being tested, including infrared and microwave technology, as well as weight sensors built in at curbside. On fully actuated signals, or semi-actuated traffic signals, pressing the button to cross a smaller side street will cause an "instant walk signal". In most states, drivers only have to wait until the pedestrian has finished crossing the half of the crosswalk that the driver is driving on, after which the driver may proceed; however, in some states, such as Utah, if the driver is in a school zone with the lights flashing, the driver must wait until the entire crosswalk is clear before he may proceed.

The U.S. state of Massachusetts allows an unusual indication variation for pedestrian movement. At signalized intersections without separate pedestrian signal heads, the traffic signals may be programmed to turn red in all directions, followed by a steady display of yellow lights simultaneously with the red indications. During this red-plus-yellow indication, the intersection is closed to vehicular traffic and pedestrians may cross, usually in whatever direction they choose (this is known as a "Barnes dance").

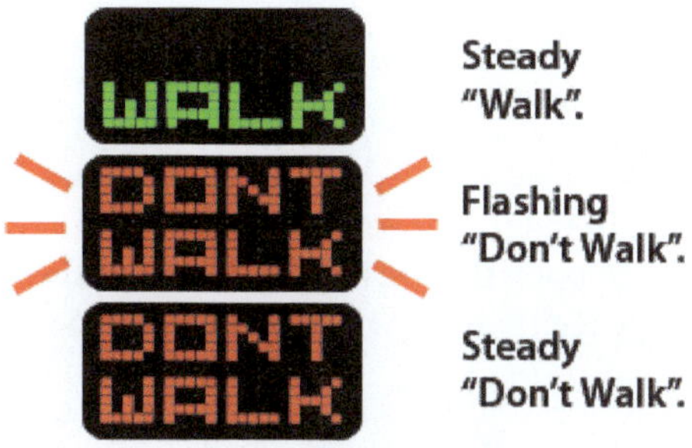

Three-state signal sequence with textual messages typical for the USA; words may be replaced by symbols.

Crosswalks have also been adapted for the blind by adding accessible pedestrian signals [APS] that include speakers at the pushbutton for each crossing location. The current standards in the US (Manual on Uniform Traffic Control Devices, 2009, Sections 4E.09 - 4E.13) require APS units to have a pushbutton locator tone, audible and vibrotactile walk indications, a tactile arrow aligned with the direction of travel on the crosswalk, and to respond to ambient sound. The pushbutton locator tone is a beep or tick, repeating at once per second, to allow people who are blind to find the device. The walk indication may be a rapid ticking tone or a speech message, accompanied by a vibrating arrow on the APS during the walk signal. APS are supposed to be set to be heard only 6 to 12 feet from the device to be easy to detect from a close distance but not so loud as to be intrusive to neighboring properties [originally many versions of audible pedestrian signals used extremely loud buzzers or even bird-like whistles to convey information, but new standards have improved upon some of these earlier devices].

United Kingdom

In the United Kingdom and certain parts of the Commonwealth of Nations, animal names are often used to distinguish several types of such crossings:

- Zebra crossing: wide longitudinal stripes on road, often with Belisha beacons; pedestrians may cross at any time; drivers must give way to pedestrians who demonstrate intent to cross.
- Pelican crossing: traffic lights for pedestrians and vehicles; button-operated.

- Puffin crossing: pedestrian lights on near side of road; button-operated with kerb-side detector.
- Toucan crossing: for bicycles as well as pedestrians.
- Pegasus crossing: an equestrian crossing.

Belisha beacons are found at zebra crossings. The other types of crossing use coloured pictogram lights, depending on the intended users of the crossing this will be a man, a bicycle or a horse.

In the United Kingdom (and in many current or former Commonwealth countries as well), it is illegal for drivers to park upon the 'zig zag' lines either side of a pedestrian crossing. This is to ensure pedestrians' view of oncoming traffic is not obscured.

A puffin crossing in the UK

Australia

In Australia, the terminology *pedestrian crossing* is used.

Pictograms are standard on all traffic light controlled crossings. Like some other countries, a flashing red sequence is used prior to steady red to clear pedestrians. Moments after, in some instances, a flashing yellow sequence (for motorists) can begin indicating that the vehicles may proceed through the crossing if safe to do so; this is fairly uncommon however. In districts with heavier traffic warranting the use of a traffic light such as inner city areas, the equivalent of the US 'standard' configuration is used.

Zebra crossings are common in low traffic areas and their approaches may be marked by zig zag lines.

Reflector signposting is used at crossings in school zones, however given that most school crossings in the country are manned, these signs only serve as a warning to motorists.

Signals

Call Buttons

The button one pushes to activate the crosswalk signal

Reports suggest that many walk buttons in some areas, such as New York City and the United Kingdom, may actually be placebo buttons designed to give pedestrians an illusion of control while the crossing signal continues its operation as programmed. In New York City, many such buttons are now rendered useless due to the automatic programming of the crosswalk, and other buttons may work, but only at certain times of day or certain periods of the year. In the United Kingdom, pressing a button at a standalone pedestrian crossing that is unconnected to a junction will turn a traffic light red immediately, but this is not necessarily the case at a junction. In Boston, Massachusetts, some busy intersections are programmed to give a pedestrian cycle during certain times of day (so pushing the button is not necessary) but at off-peak times a button push is required to get a pedestrian cycle. In neighboring Cambridge, Massachusetts, a button press is always required if a button is available, though the city prefers to build signals where no button is present and the pedestrian cycle always happens between short car cycles. In both cases the light will not turn immediately, but will wait until the next available pedestrian slot in a pre-determined rotation.

Countdown Timers

A pedestrian signal in California with countdown timer

Some pedestrian signals integrate a countdown timer, showing how many seconds are remaining for the clearing phase. The 2009 edition of the US Manual on Uniform Traffic Control Devices (MUTCD) requires countdown signals be used at all signalized intersections with pedestrian clearance intervals (flashing upraised hand phases) longer than 7 seconds.

In Mexico City, as in São Paulo, Brazil, the walking man moves his feet. In the Liberdade district (Oriental district) of São Paulo, the walking man is replaced by pictures of typical Japanese architecture.

In Taiwan there is generally no crossing manner. The majority of crossings cannot be controlled by pedestrians, although there are exceptions in Taipei. All the crossings feature animated men who will walk faster immediately before the traffic signal will change. There is also always a countdown timer.

Variations

A pedestrian crossing with a Vienna Convention standard sign indicating to motorists that they must give priority to pedestrians using it.

In some countries, instead of "don't walk", a depiction of a red man or hand indicating when not to cross, the drawing of the person crossing appears with an "X" drawn over it.

Some countries around the Baltic Sea in Scandinavia duplicate the red light. Instead of one red light, there are two which both illuminate at the same time.

In many parts of eastern Germany, the design of the crossing man (Ampelmännchen) has a hat.

Temporary Signals

In certain circumstances, there are needs to install temporary pedestrian crossing signals. The reasons may include redirecting traffics due to roadworks, closing of the permanent crossing signals due to repairs or upgrades, and establishing new pedestrian crossings for the duration of large public events.

The temporary pedestrian crossings can be integrated into portable traffic signals that may be used during the roadworks, or it can be stand-alone just to stop vehicles to allow pedestrians to safely cross the road without directing vehicle movements. When using the temporary pedestrian crossings signals for roadworks, there should be consideration on signal cycle time. The pedestrian crossing cycles may add longer delay to the traffics which may require additional planning on road work traffic flows.

Portable pedestrian crossing signal at a road work site

Depending on the duration and the nature of the temporary signals, the equipment can be installed in different way. One way is to use the permanent traffic signals mounted temporary poles such as poles in concrete-filled barrels. Another way is to use portable pedestrian crossing signals.

Enhancements for Disabled People

Tactile paving next to a crosswalk

Pedestrian controlled crossings are sometimes provided with enhanced features to assist the disabled (disabled people). Enhancements may include:

- Tactile cones near or under the control button. These rotate or shake when the pedestrian signal is green - the image of a "green man". This is for pedestrians with visual impairments.
- Tactile surfacing pattern (or tactile paving) laid flush within the adjacent footways (US: sidewalks), so that visually impaired pedestrians can locate the control box and cone device and know when they have reached the other side. In Britain, different colours of tactile paving indicate different types of crossings; yellow (referred to as buff coloured) is used at non-controlled (no signals) crossings, and red is used at controlled (signalised) locations.

- Audible signals, such as beeps or a rapid tick in order to help blind or partially sighted pedestrians; or a short recorded message, as in Scotland, Hong Kong, Singapore and some parts of Canada (moderate to large urban centres), the United States, including Michigan, Massachusetts, Maryland and Texas. San Francisco, California has one of the most comprehensive Accessible Pedestrian Signal policies in the United States, and has installed more than 1,000 audible devices. In Japan, various electronic melodies are played, often of traditional melancholic folk songs such as "Tōryanse" or "Sakura". In Croatia, Denmark and Sweden, beeps (or clicks) with long intervals in-between signifying "don't walk" mode and beeps with very short intervals signifying "walk" mode.
- A vibrating button in addition to an audible signal is used in Australia, Germany, and some parts of the United States, Greece, Ireland, and Hong Kong to assist hearing-impaired people
- Electrostatic, touch-sensitive buttons; these require no force to activate. To confirm that a request has been registered, the buttons usually emit a chirp or other sound. They also offer anti-vandalism benefits due to not including moving parts which are sometimes jammed on traditional push-button units.
- In Perth, Western Australia, an extended phase system called 'Keywalk' was developed by the Main Roads Department of Western Australia in response to concerns from disability advocates about the widening of the Albany Highway in that city in the mid-1990s. The Department felt that extending the walk phase permanently on cross streets would cause too much disruption to traffic flow on the highway and so the 'Keywalk' system was developed to allow for those who needed an extended green light phase to cross the road safely. A small electronic key adjusted the green/walk and flashing red/complete crossing phases to allow more time for the key holder to complete the crossing of the highway safely. The system was first installed at the junction of Albany Highway and Cecil Avenue. It is unclear what became of this system.

Lighting

Crosswalk with overhead lighting, and internally illuminated overhead crosswalk signs

There are two types of crosswalk lights: those that illuminate the whole crosswalk area, and warning lights.

The Illuminating Engineering Society of North America currently provides engineering design standards for highway lighting. In the USA in conventional intersections, area lighting is typically provided by pole-mounted luminaires. These systems illuminate the crosswalk as well as surrounding areas, and do not always provide enough contrast between the pedestrian and his or her background.

There have been many efforts to create lighting scenarios that offer better nighttime illumination in crosswalks. Some innovative concepts include:

Illuminating Lights

- Bollard posts containing linear light sources inside. These posts have been shown to sufficiently illuminate the pedestrian but not the background, consequently increasing contrast and improving pedestrian visibility and detection. Although this method shows promise in being incorporated into crosswalk lighting standards, more studies need to be done.
- Festooned strings of light over the top of the crosswalk.

Warning Lights

- In-pavement lighting oriented to face oncoming traffic.
- In-pavement, flashing warning lights oriented upwards (especially visible to children, the short-statured, and smombies)
- Pole-mounted, flashing warning lights (mounted similar to a traffic signal).
- Pedestrian warning signs enhanced with LED lights either within the sign face or underneath it.

In areas with heavy snowfall, using in-pavement lighting can be problematic, since snow can obscure the lights, and snowplows can damage them.

Railway Pedestrian Crossings

Finland

A railway pedestrian crossing in Jyväskylä, Finland.

Fences in the footpath approaching the crossing force pedestrians and bicycles to slow down to navigate a zigzag path, which also tends to force that user to lookout for the train.

New South Wales, Australia

Pedestrian crossings across railways may be arranged differently, such as in New South Wales. Here they consist of

- a barrier which closes when a train approaches;
- a "Red Man" light ; no light when no train approaching
- an alarm

France

In France, when a train is approaching, a red man is shown with the word STOP flashing in red (R25 signal).

United Kingdom

Most often when a footpath crosses a railway in the United Kingdom, there will be gates or stiles protecting the crossing from wildlife and livestock. In situations where there is little visibility along the railway, or the footpath is especially busy, there will also be a small set of lights with an explanatory sign. When a train approaches, the signal light will change to red and an alarm will sound until the train has cleared the crossing.

Pedestrian Cross Striping

Pedestrian cross striping machine is special equipment professionally used for paint zebra lines on the intersections or other busy road sections. Because of the characteristics of zebra cross, parallel stripes not long, the striping machine is often small hand-guide road marking machine, convenient to turn direction. The widths of stripe are 350mm, 400mm, 450mm, 500mm and 600mm. There is discrepancy between the engineer regulations in different countries. The marking shoe of pedestrian cross striping machine, which determines marking lines' width, is much wider than other pavement marking machines. Some simple marking shoe with wheels is also applied to deal the road striping.

Painting Construction

The section of road should be swept clean and kept dry. Pull a guiding line straight and fix the two ends on the ground. Then spray or brush a primer layer on the asphalt or concrete surface. The powder state thermoplastic paint needs melt into molten state for painting. Pull or push the striping machine with the guide rod along the guiding line to keep the zebra stripes straight and beautiful. Generally, the hand-push road striper is backward coating and the self-propelled type is forward coating to conform the body force exert habit.

Safety

The safety of unsignalled pedestrian or zebra crossings is somewhat contested in traffic engineering circles.

Research undertaken in New Zealand showed that a zebra crossing without other safety features on average *increases* pedestrian crashes by 28% compared to a location without crossings. However, if combined with (placed on top of) a speed table, zebra crossings were found to reduce pedestrian crashes by 80%.

A five-year U.S. study of 1000 marked crosswalks and 1000 unmarked comparison sites found that on most roads, the difference in safety performance of marked and unmarked crossings is not statistically significant, unless additional safety features are used. On multilane roads carrying over 12,000 vehicles per day, a marked crosswalk is likely to have worse safety performance than an otherwise similar unmarked location, unless safety features such as raised median refuges or pedestrian beacons are also installed. On multilane roads carrying over 15,000 vehicles per day, a marked crosswalk is likely to have worse safety performance than an unmarked location, even if raised median refuges are provided. The marking pattern had no significant effect on safety. This study only included locations where vehicle traffic was not controlled by a signal or stop sign.

Road Surface Marking

Road surface marking is any kind of device or material that is used on a road surface in order to convey official information. They can also be applied in other facilities used by vehicles to mark parking spaces or designate areas for other uses.

Road surface markings are used on paved roadways to provide guidance and information to drivers and pedestrians. Uniformity of the markings is an important factor in minimizing confusion and uncertainty about their meaning, and efforts exist to standardize such markings across borders. However, countries and areas categorize and specify road surface markings in different ways.

Road surface markings are either mechanical, non-mechanical, or temporary. They can be used to delineate traffic lanes, inform motorists and pedestrians or serve as noise generators when run across a road, or attempt to wake a sleeping driver when installed in the shoulders of a road. Road surface marking can also indicate regulation for parking and stopping.

There is continuous effort to improve the road marking system, and technological breakthroughs include adding retroreflectivity, increasing longevity, and lowering installation cost.

History

Dead Man's Curve along the Marquette–Negaunee Road in Marquette County, Michigan, shown in 1917 with its hand-painted center line.

In the United States, the first documented use of a painted center line was in 1911 along Trenton's River Road in Wayne County, Michigan. According to the state of Michigan, the idea of using a painted center line was conceived in 1911 by Edward N. Hines, the chairman of the Wayne County, Michigan, Board of Roads, after watching a leaky milk wagon leave a white trail along a road. Hines was the fifth recipient of the George S. Bartlett Award for Highway Progress, and was inducted posthumously in 1972 into the Michigan Transportation Hall of Honor for his innovation, and was honored in 2011 with the first Paul Mijksenaar Design for Function Award.

In 1917, the idea of using painted center lines on rural state highways was conceived and/or put into action in at least three states (Michigan, Oregon, and California), apparently completely independent of one another. At some point in 1917, a white highway center line was painted along "Dead Man's Curve" on what is now County Road 492 in Marquette County, Michigan, under the direction of Kenneth Ingalls Sawyer, who served as engineer-superintendent of the Marquette County Road Commission. Sawyer was inducted posthumously into the Michigan Transportation Hall of Honor in 1973.

In Oregon in April 1917, a yellow center line was painted down the center of the Columbia River Highway, between Crown Point and Multnomah Falls, at the direction of Multnomah County Sheriff's Deputy Peter Rexford. Later in 1917, the same line was continued west of Crown Point. Rexford first conceived the idea of a yellow center line in early 1917 while riding on a bus from Salem, Oregon on a dark and rainy night, and advocated it as a safety measure on the Columbia River Highway, which Rexford patrolled as a traffic officer. When Multnomah County declined to fund the project, Rexford's boss, Chief Deputy Martin T. Pratt (later elected Sheriff), paid for the paint out of his own pocket so that the center line could be painted. Rexford later described the April 1917 line as the "first yellow center line ever painted on pavement" in the United States. An article published in *The Oregonian* upon Rexford's retirement claimed that a contest with a $10,000 reward was once held to determine the originator of the highway center line, but the contest was scrapped when information from Europe revealed that ancient civilizations had used white bricks to mark the center lines of their streets.

In the fall of 1917, Dr. June McCarroll of Indio, California developed the idea of white center lines and began advocating for their use, after she was run off the road by a truck while driving along a highway that would later be incorporated into U.S. Route 99. Dr. McCarroll soon communicated her idea to the local chamber of commerce and the Riverside County Board of Supervisors, with no success. She then took it upon herself to hand-paint a white stripe down the middle of the road, thus establishing the actual width of the lane to prevent similar accidents. In 2002, a portion of Interstate 10 was designated and signed as "The Doctor June McCarroll Memorial Freeway" in her honor.

The question of which color to use for highway center lines in the United States enjoyed considerable debate and changing standards over a period of several decades. By November 1954, 47 states had adopted white as their standard color for highway centerlines, with Oregon being the last holdout to use yellow. In 1958, the U.S. Bureau of Public Roads adopted white as the standard color for the new interstate highway system. The 1971 edition of the *Manual on Uniform Traffic Control Devices*, however, mandated yellow as the standard color of center lines nationwide. The changeover to the 1971 MUTCD standards took place between 1971 and 1975, with most done by the end of 1973, so for two years drivers still had to use the old and new. Yellow was adopted because it was already the standard color of warning signs, and because it was easy to teach drivers to associate yellow lines with dividing opposing traffic and white lines with dividing traffic in the same direction. In turn, this simple mnemonic device greatly reduced head-on collisions and improved road traffic safety.

The major downside to the MUTCD white-yellow system is that yellow has slightly less contrast than white, especially at night, so for maximum contrast, bright yellow—and highly toxic—lead chromate was used to paint yellow lines through the end of the 20th century. As a result, U.S. transportation workers must take special precautions when disturbing or removing yellow lane markings.

In England, the idea of painting a center white line was first experimented with in 1921 in Sutton Coldfield, Birmingham. Following complaints by residents over reckless driving and several collisions, the Sutton Coldfield Corporation decided to paint the line on Maney Corner in the area of Maney.

In 1971, a correspondent for the *Sutton Coldfield News* wrote an article in the newspaper recalling the event.

The line was put down as an experiment as there were a lot of accidents there, even in the early days of the motor car. The experiment proved to be so successful that the whole country adopted it as a standard road safety device, and later foreign countries paint lines on their roads, as well.

In November 2014, a glow-in-the-dark bike path inspired by Vincent van Gogh's *The Starry Night* was created with luminous paint in the Netherlands with the intent to reduce urban light pollution.

Today, road markings are used to convey a range of information to the driver spanning navigational, safety and enforcement issues leading to their use in road environment understanding within advanced driver assistance systems and consideration for future use in autonomous road vehicles.

Mechanical Markers

The cat's eye, showing the iron base, rubber housing and lenses

Mechanical devices may be raised or recessed into the road surface, and either reflective or non-reflective. Most are permanent; some are movable.

- Cat's eye, invented by Percy Shaw in the 1930s, Cat's eyes equip many major routes in the British Isles. They consist of four reflective lenses mounted in a durable white rubber housing, two facing fore and two facing aft. The housing is mounted within a cast iron shoe, which the rubber housing sinks in to when driven over. This provides protection from snow ploughing and allows the lenses to be self-cleaning—they pass a rubber blade when depressed. The lenses are available in a variety of different colours, mainly white, yellow/orange, green, red, and blue.

- Botts' dots (low rounded white dots), named for the California Caltrans engineer Elbert Botts, who invented the epoxy that keeps them glued down, are one type of a mechanical non-reflective raised marker. Generally they are used to mark the edges of traffic lanes, frequently in conjunction with raised reflective markers. Botts' dots are also used across a travel lane to draw the drivers attention to the road. They are frequently used in this way to alert drivers to toll booths, school zones or other significant reduction of speed limit. They are normally only used in warm climates since snow plows usually remove them along with the snow.

Raised-profile marking machine

- Rumble strips are commonly used for the same purpose. A rumble strip can be a series of simple troughs (typically 1 cm deep and 10 cm wide) that is ground out of the asphalt. Other alternatives, similar to the Botts' dots, use raised strips, painted or glued to the surface. A specific form of raised strips using thermoplastic is called profile thermoplastic markings. The markings are created by fusing thermoplastic to the pavement and create alternating elevation and recession pattern. This can be done as inverted-profile markings or raised-profile markings. Inverted-profile markings are created by pressing a cog rolling over the markings while they are wet to make them corrugated. Raised-profile markings are created by extruding extra thickness of thermoplastic at a specific interval to create bumps. Raised-profile markings are sometime known as convex traffic lines. Uses of rumble strips can be across the travel direction (to warn of hazards ahead) or along the travel direction (to warn of hazards of not staying within a specific lane). Their main way of function is creating a strong vibration when driven over that will alert a driver to various upcoming hazards both by sound and the physical vibration of the vehicle. In some areas in the United States, these rumble strips are located along the side of county, state, and inter state highways. They are oriented so that a driver wandering off the lane of traffic would come in contact with these strips and be forewarned that they are about to leave the roadway. The idea was first started in Atlanta, as a measure to help truck drivers on the bypass after a number of accidents in the late 1980s involved drivers who became distracted or fell asleep and left the road.

White raised pavement marker near "pea-structure" side-line on highway surface

- Reflective markers are used as travel lane dividers, to mark the central reservation (median) or to mark exit slip-roads. Incorporating a raised retro-reflective element, they are typically more visible at night and in inclement weather than standard road marking lines. The color of markers varies depending on the country of use; freeways in the United States often use reflectors manufactured to appear white to drivers proceeding in the proper direction of travel, and appear red on the reverse to warn drivers that they are proceeding *against* the proper direction of travel, creating a danger of a head-on collision. Reflective markers are also referred to as raised pavement markers, road studs, and sometimes (generically) in the UK and Ireland as cat's eye, although this name refers to one particular brand of product. These markers can be used for other purposes such as marking the locations of fire hydrants (blue) or at gates of gated communities to indicate that emergency service vehicles have a code or device that allows them to open the gate. In the United Kingdom and elsewhere, raised markers are used to mark crosswalks (crossings) to assist the blind in crossing streets. In colder climates, reflective markers may be installed below ground using an elongated narrow triangle, cut into the road surface that allows the device to be installed below the road surface. Newer technology allows these to be placed above ground with snowplough-able rails that attempt to protect the reflective components from the snowplough blade.

Confusion with Marks Left by Roadwork

Typical dash marks in the middle of the lane after dowel bar retrofit roadwork

Sometimes the result of roadwork may leave some marks on the pavements. An example is the dowel bar retrofit process to reinforce concrete slabs in order to extend the life of the older concrete pavements. The completion of the process leaves a symmetrical pattern of dash marks on the roadway as if there are associated meaning to the pattern. Motorists may interpret them as an unknown form mechanical markers, or strange road surface markings when there are many of them along the roadway. In some cases, these marks create some confusions to motorists. When the roads are under construction and the lanes are shifted, those marks may interfere with lane markings. As the marks from the dowel bar retrofit are not intended to be any form of road surface markings, the responsible agencies try to make these marks less visible to the motorists as possible.

Non-mechanical Markers

Paint

Paint, sometimes with additives such as retroreflective glass beads, is generally used to mark travel lanes. It is also used to mark spaces in parking lots or special purpose spaces for disabled parking, loading zones, or time-restricted parking areas. Colors for these applications vary by locality. Paint is a low-cost marking and has been in widespread use since approximately the early 1950s.

Paint consists of three main components: pigments, resins or binders, and water or solvents. Pigments are finely grounded materials that give out colors or block out the surface beneath it. They may contain other materials such as UV stabilizer, and fillers which bring out the color pigments to the required level. Resins or binders are the glue of the paint to bind pigment and glass beads together to the road surface. The resins for the water-based paints are polyvinyl acetate latex, methyl methacrylate or acrylic resin. The resins for solvent-based paints are linseed or soya oils and alkyd resins. The pigments and resins are mixed with water for water-based paints and solvents for solvent-based paints so that they can be applied onto the road surface. Solvents that are use can be naphtha, toluene, methanol, methylene chloride, and acetone. Due to environmental concerns, some jurisdictions has some restriction on the solvent-based paints.

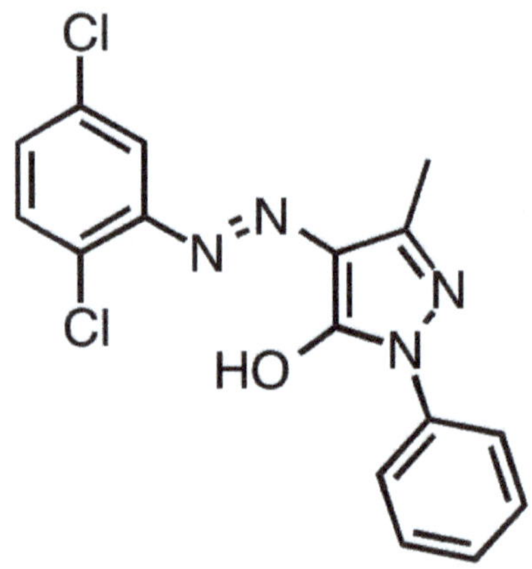

Pigment Yellow 10 is a common yellow colorant used for road marking on U.S. highways.

Paint is usually applied right after the road has been paved. The road is marked commonly by a truck called a "Striper." These trucks contain hundreds of gallons of paint stored in huge drums which sit on the bed. The markings are controlled manually or automatically by the controller who sits on the bed. Paint is run through a series of hoses under air pressure and applied to the roadway surface along with the application of glass beads for retroreflectivity. After application, the paint dries fairly quickly. Sometimes the glass beads are mixed in with the paint and apply together which is not a recommended method. The usual method is to use a separate gun to spray the glass beads on to the wet paint during the application.

Parking lot stripes and disabled symbol

Painted symbols, such as turn-lane arrows or HOV lane markers, are applied manually using stencils. Painted markings usually last 9 to 36 months. There are some water-based paints that can be applied at double the level of thickness than typical latex paints. This technique can extend the life of the markings.

Thermoplastic

One of the most common types of road marking based on its balance between cost and performance longevity, thermoplastic binder systems are generally based on one of three core chemistries: hydrocarbons, rosin esters or maleic modified rosin esters (MMRE). Thermoplastic coatings are generally homogeneous dry mixes of binder resins, plasticizers, glass beads (or other optics), pigments, and fillers. Their use has increased over paints mainly due to the performance benefits of increased durability, retro-reflectivity, and a lack of VOC solvents.

Thermoplastic markings are applied using specially designed vehicles. Usually thermoplastic marking mode should apply with equipment called road marking machine to coating traffic lines, and the road paint needs preheating by a device commonly called a preheater. The thermoplastic mix is heated in trucks to about 200 °C (400 °F) before being fed to the application apparatus. This is often a screed box or ribbon gun. Immediately after the thermoplastic has been applied, glass beads are laid onto the hot material so that they embed before the plastic hardens. These beads provide initial retroreflection. As the marking wears during use and the initial beads are lost, the beads mixed with the binder are uncovered, providing long term retroreflectivity. Most thermoplastic is produced in white and yellow colors, but other colors such as red may also be produced.

In the warm climate areas, the thermoplastic markings can last 3 to 6 years. However, snowplows can damage the thermoplastics which would limit the usage in the cold climate areas.

Preformed Thermoplastic

Preformed thermoplastic pavement markings (sometime called "tape") are thermoplastic cut into the final shapes by the manufacturers and ready to position onto an asphalt or concrete pavement surface. Preformed thermoplastics are put into place on the road surface and applied using a propane heat torch. Some models require heating the road surface prior to the placement of the preformed thermoplastics. These markings are used primarily because of their durability and cost-effective service life. As the plastics are melted into the surface, therefore they are not easily damaged by the snowplows. Typically, the preformed thermoplastic markings can last 3 to 6 years. The most common applications of preformed thermoplastic pavement markings are found at intersections as transverse markings such as stop lines, legends, crosswalks, arrows, bike lane symbols, and accessibility symbols.

Preformed thermoplastic markings ready to be applied to the road surface with a blow torch in Brussels, Belgium

Preformed Polymer Tape

Preformed polymer tapes used for crosswalk markings in a street of New York City.
Notice a tape failure on the left most stripe.

Commonly referred to as tape or cold plastic, this product is heavy-grade material with reflective beads embedded in the plastic. It is commonly used to mark crosswalks, stop bars, and traffic guidance such as turn lanes, HOV lanes, train crossings, pedestrian crossings, taxi lanes, bus lanes, and bike lanes. There are two ways to apply tape:

- Overlay—The application being laid over the surface of the pavement. Using industrial-grade rubber cement, once the tape is combined with the pavement, it should last three years. Major obstacles to estimated life are snow-plows, salt, and mis-application.

- Inlay—The tape physically becomes part of the asphalt. Using the heat generated in the paving process, road workers lay special tape on the asphalt in the hardening process, and rollers compress the two together.

The life of the preformed polymer tapes can vary based on the applications. If applied correctly, they can last between 4 and 8 years. However, there have been cases where tape failures start soon after the installation. Conditions that may contribute to the tapes to peel off are the time of year of the installation that is too close to the winter, surface preparation and workmanship. A technique that can be used to minimize the tape getting scraped off by snowplows is sandblasting a groove into the surface and fixing the tape onto this groove. This technique diminishes the advantage of low labor cost of the tapes.

Epoxy

Epoxy contains two parts which are a pigmented resin base and catalyst. The two parts are mixed in a specialized truck for epoxy marking application. The epoxy is then heated prior to spraying onto road surface. Retroreflective glass beads are applied using a separate bead gun behind the epoxy spray gun. Typically, epoxy markings last about 4 years.

Epoxy has been in use since the late 1970s and has gained popularity over the 1990s as the technology has become more affordable and reliable. This material competes directly with plastic with respect to usage and cost.

Negative Impacts to Road Surface

Road surface raveling and potholes found exclusively along pavement markings

Non-mechanical markers are found to contribute to the deterioration of asphalt concrete road surface courses. The paint and tape markers can cause the road surface to crack, and in more severe cases, the markers contribute to road surface raveling (a process in which the aggregate particles are dislodged from the road surface, causing the surface texture to become deeply pitted and very rough) or potholes.

This type of surface damage can be found exclusively underneath the pavement markings such as lane markings and turn-lane arrows. There is no definitive explanation of the relationship between pavement markings and surface deterioration, but there are several hypotheses. One is that water vapor may have been trapped underneath the road surface markings, causing the de-bonding of asphalt binder from the aggregate materials. Another hypothesis is that the reflectivity of the markings may create differences in solar heating and thermal expansion strains between the areas with and without markings. Small flaws caused by differential strains may be combined into longitudinal cracks along the markings. There are certain surface treatments that can make the road surface less susceptible to this type of distresses, such as slurry seals and stone mastic asphalt.

Temporary Markers

Pylons are sometimes used to separate HOV lanes from regular traffic lanes. They are also used in areas where lanes are used at different times for travel in both directions. These pylons have shafts that drop into holes in the road surface. A good example of this type of use was the Golden Gate Bridge in San Francisco, before they switched to a moveable barrier system.

Removable tapes can also be used in road construction sites as temporary markers. The tapes can be placed to shift the lanes and black tapes can be used to temporarily blackout the existing markings. At the end of the constructions or within 6 months, the tapes can be pulled out off the surface without using heat, solvents, or machines.

Country Specific Information

The Americas

Most countries in North and South America have yellow lines separating traffic directions. However, Chile uses white lines.

Canada

Generally speaking, Canadian pavement marking standards are consistent with those used throughout the United States.

Yellow lines are used to separate traffic moving in opposite directions, and white lines are used to separate traffic moving in the same direction, and on the shoulders of paved roads. On one-directional roads, a yellow line appears on the left shoulder, and a white line on the right shoulder. Passing rules are denoted by dashed lines as in the United States. Orange painted lines are sometimes used when the direction of the road is altered temporarily for construction projects. However, the colour scheme was reversed before 1971, when white was formerly used to denote the separation of opposing traffic, and yellow lines, when used, to denote the separation of the paved road from the right-hand shoulder. Canada adopted similar road marking standards to conform with those adopted in the USA between 1971 and 1978.

Broken lines that are wider and closer together than regular broken lines are called continuity lines. Continuity lines on the left side of a lane denote that the lane is about to end and that motorists must soon merge left. Continuity lines on the right mean that the lane will continue, but traffic may merge into it ahead.

In some areas, reflective markers (cat's eyes) recessed into the pavement are used, especially approaching curves in the road.

Ontario has several pavement marking test areas located in various parts of the province. Perhaps the most well-known location is the eastbound lanes of Highway 401 near Belleville. Other test sites are located on the westbound lanes on Highway 417, east of Ottawa, Highway 60 West of Renfrew, Highway 28 east of Bancroft, Highway 69 North of Honey Harbour and on Highway 37, South of Tweed. Pavement marking manufacturers from around the world supply a variety of materials for these sites to have their products evaluated and approved for use on provincial highways.

United States

A typical stretch of Valencia Boulevard in Valencia, California, where the lanes are marked only by Botts' dots.

In the U.S., the type, placement, and graphic standards of traffic signs, and road surfaces are legally regulated—the Federal Highway Administration's *Manual on Uniform Traffic Control Devices* is the standard, although each state produces its own manual based upon the Federal manual.

Generally white lane markings indicate a separation between lanes traveling in the same direction while yellow markings indicate opposing traffic on the other side of the line. In some areas, such as Colorado and Florida, black material is applied on the surface before a shorter white line is painted. This improves the contrast of the marking against "white" pavements, such as concrete or faded asphalt.

In California, Botts' dots are commonly used to mark lanes on most freeways. A large number of California cities also use Botts' dots on some (or all) major arterial roads. The notable exception is the city of Los Angeles which only uses paint.

In California and Nevada, the reflectors when present *are* usually the lines, and no paint is used for additional markings. Exceptions include: freeways built from white concrete where painted stripes are added to make the lanes more visible through sun glare, freeways built so wide that the risk of drifting is minimal (e.g., Interstate 5 in the Central Valley), and freeways in areas where it snows in the winter (since the snowplows would scrape off the Botts' dots).

In general, single broken lines mean passing or lane changing is allowed, single solid white lines mean lane changing is discouraged but not prohibited, and double solid white lines mean it is prohibited, as it often is in tunnels. On two-lane roads, a single broken center line means that passing is allowed in either direction, a double solid center line means passing is prohibited in both directions, and the combination of a solid line with a broken line means that passing is allowed only from the side with the broken line and prohibited from the side with the solid line. The solid white line on the right side is called the Fog Line used to help cars stay in their lane during foggy conditions and help pedestrians stay off the road

Marked crosswalks are indicated at a minimum by a pair of white lines. On major boulevards, crosswalks are further highlighted by zebra stripes, which are large white rectangles in the crosswalk perpendicular to traffic. In order to maximize the longevity of zebra crossing stripes, they are usually applied to correspond with the portions of the lane on which the wheels of a car are not usually traveling, thereby reducing wear on the markings themselves.

Pavement markings are used in addition to signs in the United States to designate regular and handicap parking spaces. Regular parking spaces are normally marked with white, although other colors are used. Handicap parking spaces are required to be located within a specified distance of an entrance. Blue markings are used to indicate the parking space is reserved for persons with a disabled parking permit.

Asia

Hong Kong

Road markings in Hong Kong is basically identical with the United Kingdom, with longer dashed white lines to indicate lanes of opposing traffic, and shorter dashed white lines for lanes in the same direction. Solid double white lines are used to indicate that drivers are not permitted to change lanes. A solid white line with a broken white line indicates that crossing the line is allowed from the lane closer to the broken line. Double solid white lines are in place in all tunnels and underpasses.

As in the UK, solid yellow lines are painted along the kerbside to indicate that no parking is allowed, with double solid yellow lines meaning no parking is allowed at any time. Zig-zag lines are used on both ends of zebra crossings. Road studs are also used as in the UK.

Japan

In Japan, white always separates traffic in the same direction or indicates traffic in the same direction can use a buffered area that is striped in crosshatch patterns such as at right turns on two-way roads since Japan is a country that has left-side driving.

White is also used on divided expressways with a solid raised center divider. Two-lane expressways where poles are the only physical barrier between opposing directions of travel always have yellow either side of the row of poles, and white is between the yellow striping and the poles.

White is also used to denote passing allowed on other two-lane roads. Yellow indicates no passing is allowed. On all roads, yellow stripes are always solid.

On expressways where there are many sharp turns and curves, seen especially in the largest cities, a yellow line indicates no passing between lanes, as follows:

- Solid yellow beside solid white: no entry permitted from the lane the stripe is next to, but passing is permitted with caution.

- Solid yellow beside broken white: passing is permitted from the side with the broken white line, but not from the side with the yellow line.
- Solid yellow line alone: passing prohibited from either lane, used on very tight curves and in tunnels.

Other markings include in the cities, destination and exit names painted in the lanes, which is done due to the very close proximity of exits, where in many cases it would be impractical to put up many overhead signs, although these are often seen approaching exits, a curved or slanted arrow points to the side of the expressway the exit will be on. A straight arrow following characters indicates the destination of the expressway.

Where a solid white line appears between lanes, passing is generally allowed but with caution.

Europe

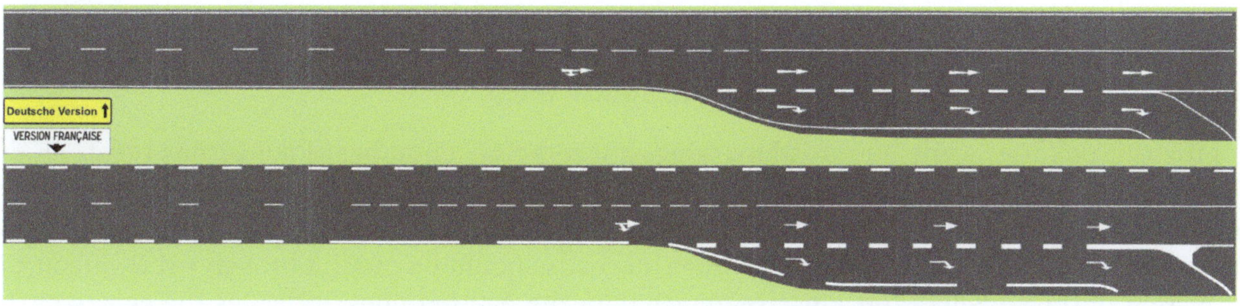

Even within the European Union every country has its own standards. Here is the same road section with German-style (at top) and French-style (below) road markings

In general European countries follow the Vienna Convention on Road Signs and Signals, which describes what road signs and road markings shall look like. The convention has some flexibility so road markings vary somewhat between the countries.

Most European countries reserve white for routine lane markings of any kind. Yellow is used to mark forbidden parking, such as on bus stops. However, for example Norway has yellow markings separating traffic directions. Many countries use yellow, orange, or red to indicate when lanes are being shifted temporarily to make room for construction projects.

In the Netherlands, Germany, Sweden, and the UK, so-called "naked roads" have been trialed, whereby all visible road markings, kerbs, traffic lights, and signs are removed, on urban roads. When this was tested in Seend, a village in the UK county of Wiltshire in 2005, the county council reported that accidents fell by a third, with motorists' speed falling by an average of 5%. It has been suggested that naked roads force drivers to make eye contact with other road users, and that it is this nonverbal communication that is responsible for the reduction of accidents. Other have suggested that road markings, especially with middle marker, make the road look like a main road, triggering faster and more relaxed driving, while no marking makes the road look like a lower quality road.

It is an experience from the introduction of marking, that at least on narrow roads and in curves, middle lines help make drivers stay on their side when meeting, and therefore reducing meeting accidents. Note that the 2005 experiment was on urban roads. In Sweden, local streets generally have no marking.

Norway

In Norway, yellow lines are used to separate traffic moving in opposite directions and on the left shoulders of paved roads, and white lines are used to separate traffic moving in the same direction, and on the right shoulders of paved roads. On roads narrower than six metres (20 ft), the center-line is removed, and the shoulder lines are broken. Short, broken lines means passing is allowed, long, broken lines means passing is allowed but dangerous, and a double yellow line means passing is forbidden. On motorways, the left shoulder is a thick yellow line, like in the US. Most other European countries use white lines for all these types of lines.

United Kingdom

During the World War II the Pedestrians Association lobbied for the government to make it safer for pedestrians to walk during the black out. As a result, white lines were painted on the sides of the road and pedestrians were allowed to use a small torch.

In the UK, the first "white line" road markings appeared on a number of dangerous bends on the London-Folkestone road at Ashford, Kent, in 1914, and during the 1920s the rise of painted lines on UK roads grew dramatically. In 1926 official guidelines were issued by the Ministry of Transport that defined where and how white lines on roads should be used. A broken white line in the direction of travel, where the gaps are longer than the painted lines, indicates the centre of the road and that there are no hazards specific to the design and layout of the road, i.e. no turnings, sharp bends ahead etc. A broken white line in which the gaps are shorter than the painted lines indicates an upcoming hazard, the proportion of white to black indicates the degree of hazard i.e. more white means more hazard.

A double solid white line indicates that the line may not be crossed, overtaking is permitted if it can be performed safely without crossing the line. Solid lines can be crossed in certain specific conditions (entering premises, overtaking a stationary vehicle, overtaking a vehicle, pedal cycle or horse travelling at less than 10 mph, or when directed to do so by a police officer). A solid white line with a broken white line parallel to it indicates that crossing the line is allowed for traffic in one direction (the side closest to the broken line) and not the other.

Solid white lines are also used to mark the outer edges of a road.

A double yellow line (commonly known as just a "Double Yellow") next to the kerb means that no parking is allowed at any time, whilst a single yellow line is used in conjunction with signs to denote that parking is restricted at certain times. Double and single red lines mean that stopping is not allowed at any time or between certain times respectively.

On many roads in the UK, retro-reflective road studs, including those known as "cat's eyes" when referring to the Halifax type road stud, are placed in the road. These devices reflect the light from a car's headlights back towards the driver in order to highlight features of the road in poor visibility or at night. The colour of road studs differs according to their location. Those defining the division between lanes are white, red road studs are placed along the hard shoulder of motorways, dual carriageways and other roads to mark the left-hand edge of a running lane; and orange road studs are placed along the edge of the central reservation. Green road studs denote slip roads at grade-separated junctions and also road-side lay-bys.

Comprehensive information about highway markings in the UK can be found in the Highway Code and on the gov.uk website.

Oceania

In areas with high levels of Snow, such as Snowy Mountains NSW, (where this picture was taken), yellow lines are used to contrast the white snow. The markers on the side of the road are also taller than other markers in NSW and pink, in order to increase visibility in snowy weather

Australia

In Australia, white lines are generally used both to separate traffic flowing in the same direction and traffic flowing in opposite directions. Double solid white center-lines may not be crossed under any circumstances, unless avoiding an obstruction. Dashed lines may be crossed for overtaking, changing lanes or turning, and also in the case of double-line markings provided the dashed line is on your side of the markings. For this reason dashed lines are usually used to mark multiple lanes traveling in the one direction. Yellow lines along road edges are used nationally to indicate "No Standing" areas not otherwise marked by signs. Solid white lines are also used to indicate kerbside parking, pedestrian and bicycle lanes, and other kerbside features. Yellow line markings are also used in areas that receive regular annual snowfall to provide contrast. Double-line markings are used to separate traffic flowing in opposite directions on busy roads.

The Bridge over the Thredbo River. One of the points in which the white lines become yellow as the amount of snowfall increases

Solid white lines are used to mark an intersection that a driver must stop at before entering whilst obeying all Right of Way laws. Dashed white lines are used to mark an intersection at which a driver must Give Way. Dashed white lines are also commonly used to indicate turns in intersections and to indicate intersections where a diamond turn is possible (intersections in which two cars traveling in opposite directions turn to the same direction-of-travel as each other without coming into contact).

Materials used are waterborne paint, thermoplastics, and cold applied plastic (PMMA), all with glass bead. Bead is generally 1mm for longitudinal marking. Currently moving to performance specified contracts with the primary performance indicator being retro-reflectivity measured with 30 metre geometry instruments. Intervention levels vary generally from 100 - 150 mcd/lux/m2.

New Zealand

Although New Zealand follows the convention of a solid yellow line to indicate no passing on roads with two-way traffic, it uses long dashed white lines to indicate when passing against opposing traffic is allowed on two-lane roads *and* shorter ones to separate lanes going in the same direction. The New Zealand convention followed the USA MUTCD convention common between 1961 and the early seventies.

References

- Day, Lance; Ian McNeil (1996). Biographical dictionary of the history of technology. Taylor & Francis. pp. 404–5. ISBN 0-415-06042-7.
- Pedestrian planning and design guide. Land Transport New Zealand / New Zealand Transport Agency. 2007. ISBN 978-0-478-30945-4.
- Lansing, Jewel & Leeson, Fred (2012). Multnomah: The Tumultuous Story of Oregon's Most Populous County. Corvallis, OR: Oregon State University Press. p. 63. ISBN 978-0-87071-665-2.
- Jones, Douglas V. (1994). The Royal Town of Sutton Coldfield: A Commemorative History. Sutton Coldfield: Westwood Press. ISBN 0-9502636-7-2.
- Federal Highway Administration (2003). Manual on Uniform Traffic Control Devices (Revision 2, 2003 ed.). Washington, DC: Federal Highway Administration. ISBN 9780935403817.
- "The world's busiest bike path is in Copenhagen - INDEX: Design to Improve Life®". INDEX: Design to Improve Life®. 2011-03-30. Retrieved 2016-06-10.
- "The best roundabout design for cyclists. The safest Dutch design described and an explanation of why this is the most suitable for adoption elsewhere". A view of the cycle path. Retrieved 9 July 2015.
- Impacts of LED Brightness, Flash Pattern, and Location for Illuminated Pedestrian Traffic Control Device, Federal Highway Administration, May 2015
- "2". Pavement Marking Handbook (PDF) (August 2004 ed.). Texas Department of Transportation. 1 August 2004. pp. 2–36—2–37. Retrieved 25 January 2015.
- "Pavement Marking Innovations Paving the Way for Progress". MTO Road Talk. Ministry of Transportation of Ontario. 10 (1). Archived from the original on January 11, 2015. Retrieved January 6, 2009.
- "Peak Retreats Alpine activities. From road cycling in a variety of resorts such as Samoens, Les Gets, Vaujany, St Sorlin". Peak Retreats. Retrieved 25 August 2014.
- "Signal-Controlled Pedestrian Facilities at Portable Traffic Signals" (PDF). Department for Transport (United Kingdom) (Traffic Advisory Leaflet). June 2011. Retrieved 19 May 2014.

- Haldane, David (July 15, 2001). "An Old Technology Gives New Life to Aging Concrete Roadways". Los Angeles Times. Retrieved March 24, 2014.
- Montebello, David (March 2000). "Cost of Pavement Marking Materials" (PDF). Synthesis Report 2010-2011. Minnesota Local Road Research Board. Retrieved April 22, 2014.
- Julian, Frank & Moler, Steve (July–August 2008). "Gaining Traction In Roadway Safety". Public Roads. Federal Highway Administration. 72 (1). Retrieved December 25, 2014.
- Staff (2000). "Chapter 5: For All Drivers, White Lines and Traffic Lanes". Road Users' Code. Hong Kong Transport Department. OCLC 45968644. Retrieved January 8, 2014.
- Holstege, Sean (11 September 2013). "Seeing double on I-17? It's a sign of safer times.". The Arizona Republic. Phoenix, Arizona, USA. Retrieved 3 December 2014.
- "Pavement Surface Condition Field Rating Manual for Asphalt Pavements" (PDF). Northwest Pavement Management Association. pp. 32–33. Retrieved 31 May 2014.
- Bevers, Cameron. "Frequently Asked Questions About Ontario Highways". The King's Highway. Self-published. Retrieved January 8, 2013.
- Castella, Tom (4 September 2013). "Does pressing the pedestrian crossing button actually do anything?". BBC News Magazine. BBC. Retrieved 23 November 2013.

Permissions

All chapters in this book are published with permission under the Creative Commons Attribution Share Alike License or equivalent. Every chapter published in this book has been scrutinized by our experts. Their significance has been extensively debated. The topics covered herein carry significant information for a comprehensive understanding. They may even be implemented as practical applications or may be referred to as a beginning point for further studies.

We would like to thank the editorial team for lending their expertise to make the book truly unique. They have played a crucial role in the development of this book. Without their invaluable contributions this book wouldn't have been possible. They have made vital efforts to compile up to date information on the varied aspects of this subject to make this book a valuable addition to the collection of many professionals and students.

This book was conceptualized with the vision of imparting up-to-date and integrated information in this field. To ensure the same, a matchless editorial board was set up. Every individual on the board went through rigorous rounds of assessment to prove their worth. After which they invested a large part of their time researching and compiling the most relevant data for our readers.

The editorial board has been involved in producing this book since its inception. They have spent rigorous hours researching and exploring the diverse topics which have resulted in the successful publishing of this book. They have passed on their knowledge of decades through this book. To expedite this challenging task, the publisher supported the team at every step. A small team of assistant editors was also appointed to further simplify the editing procedure and attain best results for the readers.

Apart from the editorial board, the designing team has also invested a significant amount of their time in understanding the subject and creating the most relevant covers. They scrutinized every image to scout for the most suitable representation of the subject and create an appropriate cover for the book.

The publishing team has been an ardent support to the editorial, designing and production team. Their endless efforts to recruit the best for this project, has resulted in the accomplishment of this book. They are a veteran in the field of academics and their pool of knowledge is as vast as their experience in printing. Their expertise and guidance has proved useful at every step. Their uncompromising quality standards have made this book an exceptional effort. Their encouragement from time to time has been an inspiration for everyone.

The publisher and the editorial board hope that this book will prove to be a valuable piece of knowledge for students, practitioners and scholars across the globe.

Index